AMERICAN HISTORY AND ITS
GEOGRAPHIC CONDITIONS

AMERICAN HISTORY

AND ITS

GEOGRAPHIC CONDITIONS

BY

ELLEN CHURCHILL SEMPLE

WITH MAPS

"So much is certain: history
lies not near but in nature"
CARL RITTER

BOSTON AND NEW YORK
HOUGHTON MIFFLIN COMPANY
The Riverside Press, Cambridge

Published October, 1903

TO MY MOTHER

WHO HAS ALWAYS GIVEN TO HER CHILDREN

AND DEMANDED FROM THEM

THE BEST

CONTENTS

LIST OF ILLUSTRATIONS

AMERICAN HISTORY AND ITS GEOGRAPHIC CONDITIONS

CHAPTER I

THE ATLANTIC STATES OF EUROPE THE DISCOVERERS AND COLONIZERS OF AMERICA

THE most important geographical fact in the past history of the United States has been their location on the Atlantic opposite Europe;[1] and the most important geographical fact in lending a distinctive character to their future history will probably be their location on the Pacific opposite Asia.

Till the end of the fifteenth century Europe was dominated by her position as part of the continental triad of the Eastern Hemisphere. In spite of her long, indented coast line and considerable maritime development, hers was mainly coastwise navigation, not oceanic; and her sea travel was everywhere supplemented on the great routes by land travel up or down river valleys and across deserts or steppes. Herself a country of the temperate zone, she sought the tropical products of the Orient. Thither the way was easy, for the Indian Ocean stretched out two welcoming arms, the Red Sea and the Persian Gulf, towards the Mediterranean, while the Tigris and Euphrates rivers rested their light finger touch on the highlands of Armenia near the basin of the Euxine. Moreover, into the Mediterranean Europe dropped a fringe of peninsulas, which gave her ample contact with its waters. Thus, for the easygoing trade of the middle ages, communication with the Orient was in general satisfactory. The Mediterranean was an open highway; and to the east, if the Suez route was temporarily blocked, more goods came by way of the Euphrates and Antioch, or by Trebizond and the Black Sea; or the silks and teas of China could be deposited at Batum, the eastern port of the Euxine, after a long journey across the highlands of Central Asia, down the valley of the Oxus, across the Caspian Sea, and along that deep furrow between the Great Caucasus and the Anti-Caucasus mountains which forms a natural highway to the Black Sea.

But wherever a water route is interrupted by a land barrier, as in the isthmuses of Panama and Suez; or where a continuous water route is constricted to a narrow passway, as in the Strait of Gibraltar, the English Channel between Calais and Dover, the Bosphorus, the Strait of Bab-el-Mandeb, and the Straits of Malacca, we find points of strategic importance which give to their masters the power to tie up the arteries of travel. Now any one who knows how to look at a map can see that the eastern coast region of the Mediterranean from Antioch to Alexandria partakes of the nature of an isthmus, while Constantinople is in a position to block the Trebizond and Batum routes. With the Saracen conquest of Egypt and Syria in the seventh and eighth centuries, European trade with the Orient was cut off

from the southern routes and was forced to rely on
the Bosphorus passway alone, until the crusades again
opened up the roads through Syria and Egypt. But
the day came when all three of these strategic positions
were held by the Ottoman Turks, a power with whom
it was fruitless either to fight or to reason. Europe
gazed longingly, despairingly towards the old avenues
to the Orient, whither her face had been set so long;
then reluctantly, at first almost hopelessly, she looked
towards the Atlantic, on which hitherto her back had
been turned, and asked the momentous question whether
the ocean could afford a route down the path of the
setting sun to the Indies and Cathay.

Thus was initiated the westward movement. It was
participated in by all the Atlantic nations of Europe
and by only these; in it the maritime powers of the
inland basins took no share. Land-locked seas have
always been the nurseries of maritime development, but
nurseries they have remained. In the fifteenth century
Europe could boast two groups of genuine sea powers,
the Italian cities in the Mediterranean and the Hanse
towns in the North Sea and the Baltic, each the product
of its geographical environment, born of the union of
great inland routes of trade with the indented coast
of a tideless sea.[2] But with the effort to discover ocean
routes to the Orient began what has been called the
Atlantic period in history, in contrast to the Mediterranean period which preceded it. Then there came upon
the scene of history other maritime powers, trained in
a severer school of seamanship, accustomed to look out
upon the landless horizon of the *Mare Tenebrosum,*

yet heirs to all the knowledge of navigation and cos-
mography their Mediterranean neighbors could give
them, stimulated and guided by Florentine and Genoese
and Venetian map makers, pilots, and sea captains.
These nations had the start in the race towards the
west; and when they realized that the new Occident
was not the old Orient, their geographical position gave
them the best opportunities the earth could offer for
wealth and dominion. Why this was so an analysis of
the geographical features of the Atlantic will make clear.

West of the Dark Continent and the Light Continent lies a long body of water whose area is only one half
that of the Pacific, though its drainage basin is more
than double that of the larger ocean. The Atlantic,
by the gigantic tendrils of its rivers, lays hold upon
the Rockies and the Andes, the plains of Central Russia
and the highlands of Abyssinia, and hence draws to
itself by mere force of gravity the trade from the heart
of the bordering continents. Its shape is approximately
that of the letter ' S,' the stricture in the middle, formed
by the approach of Cape St. Roque and the Sierra Leone
coast of Africa, dividing the ocean into two well-defined
basins. On the northern basin, which Europe faces,
nature has bestowed all her favors. Though 3600 miles
wide at its broadest part between Morocco and Florida,
between Norway and Greenland it contracts to less
than half that width and has therefore only limited
contact with the chilling waters of the Arctic Ocean.
In contrast to the smooth contour of the South Atlantic,
the northern basin has countless indentations great
and small, is bordered by vast land-locked seas, like the

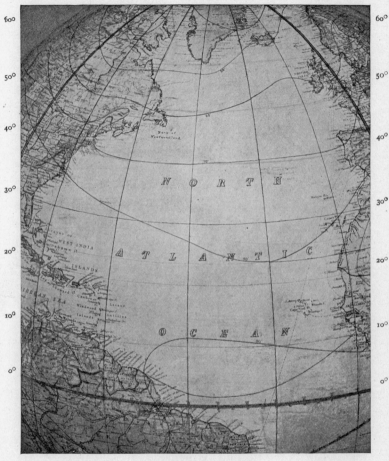

NORTH ATLANTIC BASIN

Mediterranean, the Baltic with the gulfs of Bothnia and Finland, the Gulf of Mexico, Gulf of St. Lawrence, and Hudson Bay, which carry the ocean into the interior of the continents and lay upon the bordering lands the vivifying touch of the sea.

The Atlantic as a whole has a coast-line of 55,000 miles, as long as that of the Pacific and Indian oceans taken together. This is due almost entirely to its numerous inlets, for it is island poor. The countries whose shores it washes have the ample contact with the sea which leads to a seafaring life ; but except in its great inlets, like the Mediterranean, it lacks the islands which early tempt to bold maritime enterprise. It is a significant fact that the ocean was first crossed in its narrowest part, in the only place where the islands are grouped like distant stepping-stones from shore to shore. The Northmen early became acquainted with the Hebrides, Orkneys, Shetlands, and Faroes, from which they passed to Iceland, thence to Greenland, and thence to Newfoundland or Cape Breton Island and the continent of America. Lured to the ocean by the outlying islands, bred to the sea by their deep-running fiords and their steep mountains, which denied them more than a narrow foothold upon the land, on the sea also they were forced to eke out by fisheries the slender subsistence which the country afforded. Only the small deposit plains of gravel and sand at the heads of the fiords yielded level surface for agriculture, while conditions of climate further limited the products of the soil. When we consider that Norway to-day has only 2300 square miles, or less than three per cent. of her area [3] under cultivation as

grainfields and meadow land, we can realize that in the
early centuries, with their primitive systems of hus-
bandry, population must soon have pressed upon the
limits of subsistence. It is the tendency of all people
thus placed on a narrow hem of land between mountain
and sea to establish maritime colonies, even where a
warmer climate makes nature more generous.[4] Phœnicians and Greeks dotted the Mediterranean coasts with
their settlements, and Malays to-day permeate the island
world southeast of Asia; while the Northmen planted
themselves on every available coast and island, and swept
the shores of Europe from the White Sea to the Black.
Their daring voyages were possible because their vessels
were stronger and swifter than the Spanish and Portuguese caravels of five centuries later. In this sea-born
genius for shipbuilding they suggest the Haida Indians
of the Queen Charlotte Islands, who make their canoes
on what a sailor would call beautiful lines, and in them
venture out a thousand miles into the open Pacific.[5]

But the colonizing enterprises of the Northmen in the
ninth and tenth centuries were too numerous and scattered, the climatic conditions of Vinland too harsh for
a small and remote settlement in the midst of hostile
Indians, and the population of the mother country too
sparse to produce any lasting results from these early
transoceanic voyages. Hence the effective discovery of
America had to wait for five centuries, to be made
by nations further south, to whom better conditions of
climate and soil had given larger populations, and who
also enjoyed ample contact with the sea, but who faced
the broadest stretch of the Atlantic and made their

voyages into the vast unknown. But they too were led across the threshold of the ocean by a few outlying islands. The nearer groups of the Madeira and Canaries had been known for centuries, though we hear of an expedition sent out in 1341 by Alphonso IV. of Portugal to rediscover the Fortunate Islands (Canaries). The Azores, a thousand miles west of Portugal, seem to have been visited first in 1147 by an Arabian navigator, who set out on his voyage of discovery from the port of Lisbon, at that time the Moorish stronghold of power and commerce. From 1346 they appear upon the maps of Europe, and seem to have had chance visitors from that time till their so-called rediscovery in 1431 by one of the expeditions of Prince Henry of Portugal. The long voyages to all these remote islands educated the timid sailor of the fifteenth century in maritime enterprise. Only the group of the Canaries belonged to Spain, but it was fortunately situated within the belt of the northeast trade winds, and on the twenty-eighth parallel, which happened to be also the latitude of the northern end of Cipango, according to its supposed location in the western Atlantic as set down in Toscanelli's map. Since the lack of chronometers in the fifteenth century made the calculation of longitude very inaccurate, it was the custom for navigators to sail due north or south to the parallel of their proposed destination and then shape their course directly east or west.[6] Thus the Canaries were the first objective of Columbus in his outward voyage, and, when he steered thence, served to determine the Bahamas as the first landing-place in the New World. This island outpost and the beneficent

trade wind combined to make the voyage into the
unknown of as short duration as possible for the timid
and mutinous spirits of the Spanish crew.

The Cape Verde Islands, ten degrees further south,
formed the turn in the road in the third voyage of
Columbus, whence his southwesterly course brought him
into all the horrors of the equatorial belt of calms.
These islands, a Portuguese possession, served to direct
also the course of the Portuguese navigator Cabral,
when, standing further out to sea than was customary
for vessels following the Cape route to India, and drift-
ing or driven further westward than he realized, he came
to the coast of South America, which here lies directly
in the path of the southern equatorial current, and only
ten degrees west of the longitude of the Cape Verde
Islands.   This protrusion of the South American continent forty degrees east of the Bahamas, and therefore
east of the " Line of Demarcation " laid down by the
pope, Alexander VI., threw Brazil to Portugal and gave
that power its sole domain in the New World.

That Portugal should have been the pioneer in the
search for the " outside " passage to the Indies and
should have followed the southern route was intimately
connected with her geographical location.   Through
the Mediterranean she had felt the influence of the
Oriental trade for centuries; by her outpost position
on the ocean and her long coast-line facing west and
south she was dominated by the Atlantic; while the
proximity of Africa and the outlying islands to the
southwest, rediscovered by her, naturally attracted her
maritime enterprise.   The resolution of all these geo-

graphical forces was the search for the African sea route to the Orient. Minor forces also contributed to this end. Portugal, like Aragon and Old Castile, being remote from the center of Saracen dominion in southeastern Spain, was held by a weaker grasp and was therefore able to throw it off sooner. Lisbon was liberated in 1341, a century and a half before Spain had accomplished the reconquest of her territory, and therefore could turn to other tasks. Moreover, her Moorish neighbors helped to develop the seamanship of Portugal in naval warfare with Mussulman pirates, in pursuit of whom Portuguese vessels became accustomed to sail a little way down the western coast of Africa. Afterwards such voyages were continued for commerce, just as the sequel to Prince Henry's military expedition into Morocco, where he learned of the gold-laden caravans from the Guinea coast to Mediterranean ports, was his endeavor to reach the Gold Coast by sea. From that time, in his sea-bound castle of Sagres where the promontory of St. Vincent carries Portugal territory far out into the ocean, Henry the Navigator, watching his ships outbound from the neighboring harbor of Lagos, gazing out into the Atlantic for the return of his storm-beaten caravels, is symbolic of Europe in the fifteenth century.

From Lisbon, Lagos, and Sagres the fever of maritime discovery spread along the coast to Palos, Cadiz, and Seville. Many men, like Christopher Columbus, his brother Bartholomew, and Magellan, learned much in the service of Portugal before they made their great discoveries for Spain. Columbus's sojourn in Lisbon, at that time the center of nautical science in Europe, strengthened his convictions of the feasibility of his plan to sail west. Added impulse was given his purpose in 1471 by the expedition of Santarem and Escobar which rounded the elbow of the Guinea Gulf and disclosed the southern trend of the African coast. This information occasioned probably the simultaneous inquiry of Columbus and King Alphonso V. of Portugal to Toscanelli in regard to a shorter transoceanic route to the Indies.[7] Though the southern route was sure to be long, Portugal was committed to it, and Columbus turned to Spain, where Portuguese achievement had aroused maritime enthusiasm and the spirit of rivalry which would welcome the strange proposition. And then, on the southwestern coast of Spain, in the ports of Cadiz, Palos, and San Lucar de Barrameda, was concentrated the life of the country for the next hundred years.

The safety and ease of the first route of Columbus, the mistaken idea that the Spice Isles were still to be found in the tropical regions of the western Atlantic, and finally the search for a possible water passage through the continental barrier, combined to hold the Spanish in the Gulf of Mexico and the Caribbean Sea, especially when Mexico and Peru were found to yield gold. Their explorations, therefore, extended around these two bodies of water, with some sporadic voyages further north up the east Atlantic coast, while their conquests spread from Mexico down to Chile. The marked southwestward trend of the South American coast below the Tropic of Capricorn made it evident that this part of the country, according to the papal bull, must belong to Spain. The Spanish, acting on this pope-given right, established settlements at the mouth of the La Plata River, because this stream afforded an approach from the Atlantic side to the Potosi mines on the Bolivian Plateau, and they wished to secure it against foreign attack.

England, the next Atlantic power of Europe to take part in the westward movement, had its maritime development predetermined by its geographical conditions. Its long coast-line and abundant harbor facilities encouraged the nautical proclivities of its population, which was made up of oversea elements, — Anglo-Saxons, Danes, and Norman French. As an island, its earliest means of defensive warfare lay in the navy of King Alfred. Later its merchant vessels found a near market in the ports of Norway, and its fishing-smacks drew in their nets off the coast of Iceland. Looking towards the western waters was Bristol, at the end of the fifteenth century the chief seaport of the island. Hither came the Genoese Cabot, and hence sailed the first transatlantic expedition of the English. Cabot's instructions excluded the southern course, probably to avoid encroaching on the maritime field of Spain and Portugal; so he sailed directly west, and came to the coast of Labrador or Cape Breton. His course lay westward approximately along the fifty-fourth parallel, where the distance from shore to shore, in consequence of the eastward projection of North America and the general convergence of the continents towards the north, is only about two thousand miles. But while his voyage was little over half as long as that of Columbus, the prevailing currents of ocean and air were against him in the outward passage.

Exploration of North America in the higher latitudes lay near to England as a result of her own northern location, just as in the case of the Norse discoveries; but the early English voyages, like those of the Vikings, were not followed up by European explorers because of the harsh climatic conditions of the bleak coasts thus reached. At a later date, however, the search for the Northwest Passage, when it came to be limited to Arctic regions, fell naturally to English navigators, such as Frobisher, Davis, Baffin, and Hudson. Thus was established England's claim to the sub-polar regions of the continent; but it remained barren till 1670, when Prince Rupert secured a grant of "the sole trade in all those seas, straits, bays, rivers, lakes, creeks, and sounds lying within the entrance of Hudson's straits," with all the lands bordering upon the same. This was the beginning of the Hudson Bay Company, whose traders, operating among the Indian tribes of the interior, came into collision with the French voyageurs, who were pushing up from the St. Lawrence by way of the Saguenay for the northern furs.

The English, however, had met the French before in the New World on the Banks of Newfoundland. From northwestern France two peninsulas of ragged contour project into the Atlantic and the Channel; and from the days of Cæsar and the famous *Veneti*, Brittany and Normandy have nourished a hardy maritime race who, like the Norse, have been forced to rely in part upon the sea for their means of subsistence.

These were the men, born to the sea, who first sailed from France to the western continent. Opposite Brittany and in the same latitude is southeastern Newfoundland, which marks the point of North America's closest approach to Europe south of Greenland. To the Banks of Newfoundland came these Breton and Norman fishermen as early as 1504; they were followed by the Portuguese, attracted thither by the possibility that this land might lie east of the papal meridian, and at a later date by the English, the original discoverers; for fishing-grounds are always the great cosmopolitan areas of the earth. The attempted settlement on Sable Island by the French in 1518, the exploration of the St. Lawrence soon after, and another short-lived colony near the present site of Quebec, all show that the French took this field for their own. Their activities were in effect limited to Nova Scotia, the islands about, and the Gulf and River St. Lawrence, though French names along the coast of Maine still bear evidence to their wider claims of territory, based upon voyages of exploration as far south as Cape Cod.

Until near the end of the sixteenth century we find the range of English exploration in the New World south of the Norse, north of the French, and far to the north of the Spanish, reproducing the geographical stratification of these nations in Europe. The Canary Islands and the northeast trade winds, as we have seen, conspired to draw Spanish activities somewhat further south than the Iberian peninsula, and to keep them limited in America between the equator and 30° north latitude, the parallel of St. Augustine. But from that point north to the mouth of the Penobscot River the coast remained debatable ground. Both French and English came hither, drawn out of their natural path at different times, but by the same motive, — the prizes which Spanish treasure ships afforded in time of war. There is nothing like naval warfare to turn the political geography of the earth topsy-turvy. When war broke out between France and Spain in 1521, French cruisers were attracted southwestward to make depredations on Spanish caravels as they set out from the Mexican ports; and one of these cruisers under Verrazano, after having relieved a galleon of its rich cargo not far from its home port, set westward from the Madeira Islands to make discoveries for the benefit of France, and explored the American coast from Cape Fear to the Piscataqua River. But no permanent results followed from this venture.

The first English navigators to appear in the tropical waters of the western Atlantic were the slave traders, drawn thither to find a market among the Spanish colonies for their human merchandise. Sir John Hawkins was bringing slaves from the Guinea coast to meet the steady demand in this region from 1562 to 1567; but his traffic was interrupted soon afterwards by the outbreak of war with Philip II. of Spain. Then English privateers began to wait for Spanish vessels in the western Atlantic, and thus became accustomed to the southern route and to these remote coasts. But the English had a motive leading them to combine with their naval and commercial expeditions the more substantial colonial ventures. The loss of Calais in 1557 ended England's hope of continental territory; it was a blow to her "balance of power" idea all the more serious in view of the vast dominions of Spain and the aggressive territorial policy of France. The loss might be made good in the New World. The war was scarcely begun when, in 1578, Sir Humphrey Gilbert sailed with his seven ships from Plymouth to found a colony in that problematical land of "Norumbega," which, however, he never reached. The year 1584 saw the birth of Raleigh's design to "plant an English nation in America," while at home was found in abundance the raw material out of which colonies could be made, a superfluous population induced recently by retrograde economic measures, and persistently by the limited area of the little island kingdom. For a long time and with few exceptions the colonial expeditions from England followed the southern route by way of the Canary Islands on their outbound voyages, although it was long and tedious. The reason is to be sought either in the help given by the northeast trade winds in the southern latitudes, and the difficulty presented by the prevailing westerlies further north; or in the preference of the early navigators for the route made familiar by their adventures on the Spanish main.

Thus, separated by almost a century from the explorations of the Cabots, was initiated the second phase of England's oversea enterprises; and in a hundred and fifty years her permanent settlements occupied a narrow strip of coast almost continuously from Georgia to Nova Scotia. As both extremities of this line overlapped the claims of their neighbors, the English came into conflict with the Spanish on the south, where St. Augustine and Savannah were alternate objects of attack; and with the French on the north, where the settlements of both nations along the Maine coast suffered from the mutual hostilities. Within this area the English had to suppress the colonial ventures of the Dutch, the last of the Atlantic powers of Europe to take part in the westward movement.

In the events establishing the colonial empire of Holland, geographic conditions were the leading factors. By its location at the northern end of the great trade route up the Rhone and Rhine valleys from the Mediterranean, this country, through the Hanseatic League, became the distributing-point in northern Europe for oriental products. When Portugal got control of the Asiatic trade, the commercial dealings of the Dutch were transferred from Venice and Genoa to Lisbon. But when the union of Spain and Portugal under Holland's hereditary enemy, Philip II., excluded Dutch vessels from the mouth of the Tagus, the merchantmen of the Rhine began to contend for a share in the land of the East Indies. The long coast-line of Holland, its numerous harbors, rich and active seaports, readier access to the Atlantic, and previous experience in the open ocean gained in the long voyages to Portugal, gave to the Dutch cities a broader horizon than the inclosed basin of the Baltic afforded the German members of the Hanseatic League, and hence fitted them for wider maritime enterprise.

Not till the end of the sixteenth century, therefore, did the Dutch appear upon the high seas. Henry

Hudson's discovery of Hudson River in 1609 and his report on its opportunity for fur trading led them to appropriate Manhattan Island at the mouth of the stream and the great angle of the American coast between Delaware Bay and the Connecticut River. Meanwhile the English colonies were spreading along two natural geographical lines which converged upon the Dutch holdings. From the east, the New England settlements were rapidly throwing out one frontier station after another along the protected waterway of Long Island Sound, so that the year 1639 saw an unbroken line from Newport to Stamford and the modest beginnings of Greenwich, which was within thirty miles of New Amsterdam. From the south, Virginia and Maryland colonists were spreading up towards the head of Chesapeake Bay, where it approaches within twenty miles of the mouth of Delaware River. From his trading-post on Kent Island, the Manhattan of upper Chesapeake Bay, William Claiborne was pushing up the Susquehanna River as early as 1631 to compete with the Dutch for the peltries of the Northwest. Moreover, the Dutch holdings occupied territory included in the original Virginia grant. The English were aware of the commercial and strategic value of the Hudson, as well as the danger attached to the separation of their two areas of colonization by a foreign power. So the fate of New Netherland was sealed.

Geographical conditions made the acquisition of New Netherland the most important political event up to this time in the history of the English colonies in America. The natural highway of the St. Lawrence and the Great

Lakes brought the French up to the back doors of New York and Pennsylvania, the two colonies carved out of the Dutch territory. From the head waters of the Ohio to those of the Hudson, French and English frontiers met. Lake Champlain with the Richelieu River to the north, and the Mohawk with Lake Ontario to the northwest, made two open, easy thoroughfares for army or trader between Canada and New York. Here no forest wilderness intervened to form a barrier, as it did between the New England settlements and the lower St. Lawrence. Contact was therefore immediate and, in consequence of the northwestern fur trade, of vital importance. Conflict was certain. Against the hostilities of the French, New York became the bulwark of the colonies, whom she helped to unite by demanding their aid against the common enemy. Thus she became the keystone in the arch of the English settlements, which was made all the stronger by the pressure of French aggression from above.

NOTES TO CHAPTER I

1. Ratzel, Politische Geographie der Vereinigten Staaten, p. 13. 1893.
2. E. C. Semple, Development of the Hanse Towns in Relation to their Geographical Environment, Bulletin of the American Geographical Society, no. 3. 1899.
3. Norway, Official Publication for the Paris Exposition, p. 308. Christiania, 1900.
4. Ratzel, Anthropogeographie, p. 290. Second edition, 1899.
5. Indians of Southwestern Alaska in Relation to their Environment, Journal of School Geography, June, 1898.
6. Fiske, Discovery of America, vol. i. p. 315.
7. Ibid. p. 355.

CHAPTER II

THE RIVERS OF NORTH AMERICA IN EARLY EXPLORATION AND SETTLEMENT

THE early European navigators did not find their voyages of discovery limited by the coast of North America; back of the sea-like bays, back of the bay-like estuaries, they met the tide-mingled flood of the great American rivers. Sea navigation passes without break into river navigation. Nature draws no sharp line between stream and ocean. The tide moves up the Potomac almost as far as Washington, and the waters of the Hudson at Newburgh are brackish; while the fresh, muddy current of the Orinoco is carried fifty miles out to sea. In recent years size has differentiated ocean vessels from river steamers, though deep canals are fast extending the landward reach of the sea and wiping out this distinction; but the small craft of the early navigators dropped their anchors and their barnacles into the broad, quiet flood of the American rivers, beyond the reach of storm and tide.

Rivers have always been the great intermediaries between land and sea, for in the ocean all find their common destination. Alexander's discovery of the Indus led by almost inevitable sequence to the rediscovery of the Eastern sea route. The St. Lawrence brought the French from the Atlantic into the Great

Lake basin of North America, and the hardy explorers had scarcely launched their canoes upon the tributary waters of the upper Mississippi before they were planting the flag of France on the Gulf of Mexico at its mouth.

The accessibility of continents is determined by the navigability of their rivers, and this in turn is a matter of continental build. From the huge, monotonous hulk of the African plateau the rivers come plunging down the steep escarpment within a few miles of the narrow plain of the coast; only where the Nile leaves its last terrace three hundred miles inland, has the Dark Continent been open to the light. The vast intaglios of the Euphrates, Indus, and Ganges valleys have served as highways into southern Asia. The structure of the Americas is characterized by long mountain ranges connected by trough-like valleys, continental in size and drained by streams which in flood-time look like inlets of the sea. Into the Atlantic most of these mighty currents are poured, affording waterways from the ocean into the heart of the continents. The chief Atlantic streams of North America are found between 25° and 50° north latitude, in the most desirable part of the temperate zone. Furthermore, North America presents to the Atlantic a long, low, coastal plain similar to that of Europe; and down its gentle slope pour the easy-going currents of its drainage rivers. Only in south-eastern New England does a narrowing of the lowland belt shorten the course and hasten the current of the streams, thus destroying their value for navigation. From the mouth of the St. Lawrence, therefore, to the mouth of the Rio Grande the eastern side of North America invited exploration.

During the sixteenth and the early part of the seventeenth century there are to be distinguished two great maritime movements along this shore of the continent, the fisheries which held to the coast, and the search for the Northwest Passage, which penetrated into the interior.[1] Both were profoundly influenced by geographical conditions. From Cape Cod to Cape Race natural conditions combined to stimulate the fisheries industries. The Labrador current, which sweeps around this angle of the continent, is the natural habitat of the cod, the mackerel, and the herring. The neighboring coast, fringed with fiords and dotted with outlying islands, afforded harbors and fishing-stations in abundance. In the land itself, poor soil and bleak climate refused to the early colonist more than a niggardly subsistence, but the sturdy pine forests, toughened by cold and nourished by rain, put into his hand the material for his fishing-smack. A line from Quebec to Boston defines the base of the peninsula forming the back country to most of this coast. It is a limited area, drained from north to south where its dimension is least. The rivers are therefore small, and hence did not adequately open up the back country for the fur trade, that great resource of the American pioneer. Thus the supply of peltries which each stream could command was soon exhausted.

The search for the Northwest Passage took the explorers into every inlet and up every river. These men came from a continent less than half as large as North America, and their eyes were trained to the smaller dimensions of Europe.[2] In the prevailing estuaries of American rivers, with their salty tide, they saw possible

channels which, like the Strait of Gibraltar and the Bosphorus, might lead to an ulterior sea. A man accustomed to the small streams of England might well
explore the broad, sluggish course of the James River,
as Newport did in 1608, to find a passage to a western
ocean; or, like Henry Hudson, might easily mistake the
lower Hudson for a pelagic inlet. The search for the
Northwest Passage, therefore, made the early navigators
familiar with the interior waters; but as the futility of
their quest became apparent, the knowledge acquired in
the process was of great value. If an oceanic waterway
was not to be found in the interior, furs were. It is
surprising with what rapidity the desire of the Europeans
moved from the Passage to peltries.

The best and most abundant furs were to be reached
only up the channels of the rivers; for the indiscrim-
inate killing of animals, young as well as old, female as
well as male, and the growing wariness of the creatures
made the fur fields rapidly recede from the coast. Rivers
were the only highways into the interior, and up the
rivers, preferably those flowing from the north, went the
American colonist, — the trapper, trader, and voyageur
always in advance. The history of every colony testifies
to this fact. In 1627 William Claiborne, of Jamestown,
had begun to trade up the Potomac and Susquehanna
rivers. The Dutch came to America for the sake of
the peltries, and up the courses of the rivers they were
drawn. Landing first on Manhattan Island at the
mouth of the Hudson, they had a small settlement there
in 1613 to control this highway into the back country.
Two years later they had built a post on a small

island just below Albany; by 1623, Fort Aurania was constructed on the mainland near by; and 1661 saw the beginning of Schenectady at the rapids of the Mohawk. The distribution of their other settlements betrays the same motive and the same characteristics: they were primarily trading-posts, were limited to the other two great rivers of the vicinity, the Delaware and the Connecticut, and were located well upstream. At Fort Nassau, near where Philadelphia now stands, the Dutch traders gathered in the furs of the Schuylkill and the middle and upper Delaware; Fort Good Hope on the Connecticut, near the present site of Hartford, enabled them to meet halfway the beaver-laden canoes of their savage customers. But the Dutch were shortsighted enough not to secure to themselves the natural commanding positions at the entrance of these streams, as they did in the case of the Hudson; so when John Winthrop, Jr., of Boston, in 1635 built Fort Saybrook at the mouth of the Connecticut, they were shut out of that river and forced to relinquish their upstream post. But when later a small colony of Swedes appropriated the western shore of the Delaware estuary, the Dutch, now grown wiser, overcame them and annexed their territory.

The Plymouth colony in 1634 had established a large trade with the Indians of the Connecticut and Kennebec rivers. But, owing to the smallness of the New England territory and the lack of great rivers, furs soon became scarce along the coast, and the traffic retreated rapidly up the few considerable streams, where the business was ruined by concentration of the traders

and over-competition. Broad areas, widely scattered
and remote posts, belong to the nature of the fur trade.
Springfield, a frontier settlement on the Connecticut
founded in 1636, was for a short time one of the best
points for beaver, but by 1645 the market was spoiled
by over-competition;[3] and Sir Richard Saltonstall and
others endeavored to secure a monopoly of the right to
establish trading-posts further north than those already
existing, while the Springfield colonists were forced to
resort to the duller and less lucrative occupation of
agriculture. With the exception of the Connecticut,
therefore, which added fertile meadow lands to the
attraction of the fur trade, the streams of New England,
in consequence of their limited basins and rapid-broken
courses, scarcely affected early settlement; while the
larger rivers of the country carried far into the interior
the vessel of the explorer, the canoe of the trader, and
the shallop of the colonist. At Albany and Montreal,
on the Hudson and the St. Lawrence, the traffic in peltries finally concentrated, and with far-reaching political results.

In Canada the search for the Northwest Passage was
pursued longer. The only great waterway in the temperate zone westward into the heart of the continent is
afforded by the St. Lawrence and the chain of the Great
Lakes. Here the hope of reaching the fabled " Sea of
Verrazano " had more to feed upon because of the very
length of the magnificent waterway already discovered.
Therefore, nearly sixty years after the English Hudson
and Baffin were limiting their search for the Northwest
Passage to Arctic seas, where alone it was to be found,

the problem of a water route to Cathay was luring La Salle to the head streams of the Ohio and Mississippi. It is a significant fact that all exploration and colonization in North America has been westward from its "Europe-fronting shore." Even the Mississippi, though first reached near its mouth by the Spanish, was effectively approached and appropriated through its westward flowing tributaries. Moreover, this approach which gave possession was made in its northern course by the Wisconsin River, at a point whence the distance east to the Atlantic shore is greatest.) Not the wintry silence of the frozen north, nor river rapids, nor storm-swept lakes, nor swamp-covered portages, nor two thousand miles of travel sufficed to check the westward moving tide.

Close on the heels of the French explorer came the voyageur, drawn landward by the magnet of the fur trade. Traffic in peltries sprang up along the shores of Acadia side by side with the fisheries; but in Canada, as elsewhere in North America, the presence of a big river determined its progress into the interior. Moreover, the location of this stream to the north, and its connection with the Great Lakes bordering on the vast forests of the bleak Northwest, gave the French access to a field yielding furs of finer quality and longer staple than any before known; while the remoteness, vastness, and harsh climate of the region promised to insure it against excessive exploitation. An infant colony, in order to survive, must command some source of large profits, such as tobacco yielded to the Chesapeake Bay settlements, the fisheries to New England, sugar and spirits to the West Indies. The reason is to be found in the remoteness of their market, the increased cost of such commodities as they buy in exchange, and the necessity of some indubitable allurement, like the prospect of wealth, in the new country to tempt settlers to make the long voyage from their native land and face the hardships of the wilderness. The glaciated soil of Eastern Canada could promise no great fertility, its bleak climate no luxuriant vegetation; but there was money in peltries. Hence the upstream advance of trading-post and settlement went on with amazing rapidity. Significantly enough, the long stretch of the St. Lawrence gulf and estuary was left almost unmarred by human habitation. Only where the gloomy gorge of the Saguenay River opened up an avenue northwestward to the Hudson Bay country and its rich furs did the trading-post of Tadousac break the solitude of the lower St. Lawrence.

The first settlements were made at the mouth of the river in 1608, at Quebec, Beaupré, Beauport, and the Island of Orleans. By 1670 there were missions, each the center of a busy trade, at the outlets of Lakes Superior and Michigan, and farther west at the head of Green Bay. As the waterways determined the landward stretch of the French claim, so they fixed the proportions of the *habitant's* land grant and the location of his cabin. His farm was always on the river, which served as the highroad. River frontage was in demand, therefore, so his land grant was very narrow, measuring its width by the foot and its depth by the mile. His dwelling faced the stream, and hence was close to that of his neighbors. Thus the settlements were strung out in long, thin lines on the margins of the rivers. "One could have seen nearly every house in Canada by paddling a canoe up the St. Lawrence and Richelieu."[4] The riparian grouping of the colonists, though eminently convenient for the Indian trade, made defense and government difficult. Like all natural forces, this geographical control of the rivers was so strong that it nullified royal decrees designed to concentrate the inhabitants and form villages.

Missionary zeal and dreams of far-reaching empire were factors in the westward movement of the French; the fur trade, on the other hand, was a determinant, because it became the sole support of settlement and fort in all the wide region of Canada. But the landward expansion of commerce and church and state was permitted and controlled by the Canadian lakes and rivers. The recurrent westward search for the problematical "great water" of the Indian anticipated the trader as it had done in the beginning. Then the remotest trading-post became a base for the next outward expedition of the explorer. And the passage from the Lakes to the Mississippi was made almost as easily as that from the Atlantic to the Lakes. This was the result of geographical conditions.

The mighty trough which runs through the center of the North American continent, from the Arctic Ocean to the Gulf of Mexico, is met halfway between these points by a lesser valley to the eastward occupied by the Great Lakes. The rim dividing these two basins is low and narrow, running close to the margin of the

Lakes. West of Superior and Michigan the ill-defined watershed is flooded in a wet season; and at all times the ponds and pools and the endless meanderings of the connecting streams tell of a drainage system unfinished because of too slight incline. Hence [the transition from the Lakes to the small affluents of the Mississippi tributaries was easy, especially as the French early adopted the canoe of the Indian as eminently suited to the low, narrow divides of the region. In flood-time there was an unbroken waterway; but the rule was to paddle up a short stream flowing into one of the Great Lakes, carry the canoe over a portage one to ten miles in length, and then launch it on a stream beyond, to float down to the all-receiving bosom of the Mississippi. The great number of the portages both offered a choice to the traders, especially advantageous in time of war with one or the other of the Indian tribes inhabiting this region, and enormously increased the voyageur's sphere of activity by opening up to him all the country drained by the Ohio and Mississippi.]

The order in which these portages were discovered[5] and used by the French was determined in part by the political neighborhood of the hostile Iroquois along the southern shores of the nearer lakes, and in part by geographical conditions. Up to 1670 the westward route of the canoe fleets was up Lakes Ontario, Erie, and Huron, to the westernmost outpost at Michilli-mackinac on the Strait of Mackinaw. This trading-station could be reached also by a more direct northern route from Montreal up the Ottawa River, across to Lake Nipissing and Georgian Bay,—a route which was

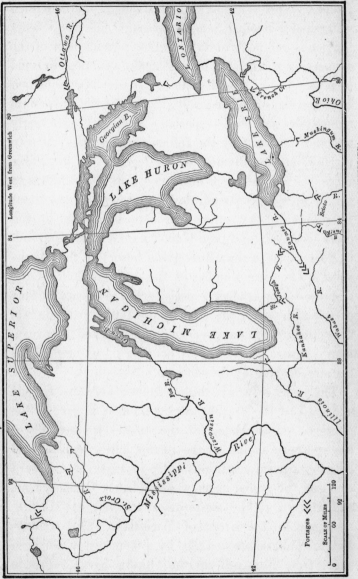

PORTAGES BETWEEN THE GREAT LAKES AND THE MISSISSIPPI

often used when the Iroquois were on the war-path about
Ontario and Erie. Further progress thence naturally
led up the waterways afforded by Lake Superior and
Lake Michigan to the portages which the Indian described as leading to the western waters. The consequence was that the passages earliest known were those
remotest from the Canadian settlements. The first carry
route used by the French led from the head of Green
Bay up Fox River, across the northern part of Lake
Winnebago, and along the upper Fox River for sixty
or seventy miles to a narrow portage of two miles, leading to the head of the Wisconsin River.[6] This was the
route followed by Marquette and the fur-trader Joliet
in 1673, when they explored the Mississippi to the mouth
of the Arkansas. The other of these earlier portages
was across the narrow, lake-dotted watershed between
the St. Louis River, which flows into the western
extremity of Lake Superior, and the easternmost bend of
the Mississippi.

By the end of the seventeenth century the portages
at the head of Lake Michigan were the best known.
There were a number of these close together. The
Chicago River, the lesser and greater Calumet, and the
St. Joseph all led by short carries to the Des Plaines
and Kankakee branches of the Illinois. The Lake Erie
portages, though affording nearer, more direct routes
by the Ohio tributaries to the Mississippi, came into use
only in the eighteenth century, and here, too, the progress was from west to east. When the French from
their base in the newly founded Detroit (1701) were able
to hold the neighboring Iroquois in check, the Maumee

River with its portages west to the St. Joseph and the Wabash was open to travel; afterwards the portages from Lake Erie to the Big Miami, Scioto, and Muskingum were rendered safe. When in the middle of the eighteenth century the French began to draw their lines closer around the back of the English settlements, that from Presque Isle (site of Erie) to French Creek and the Allegheny River marked the entrance to the fortified route of the French to their post at the forks of the Ohio, where Pittsburg now stands.

Having swept westward up the broad highway of the Great Lakes to the upper reaches of the Ohio and Mississippi, the French went to work more slowly to secure all the portages to these rivers and to occupy every strategic point at the extremities of the Lakes. Forts Frontenac (Kingston), Niagara, Rouillé (Toronto), and Detroit were designed to command the Lakes, as Fort St. Louis on the upper Illinois and the post of Outianon on the upper Wabash controlled the Chicago and Maumee portages; while Vincennes, Cahokia, and Kaskaskia completed the line of connection with the lower Mississippi. The vast area thus opened up to the French by their possession of the two great waterways of the continent led to the ultimate shipwreck of their colonial venture, because the large extent of the territory lured them to exploit its trade rather than make permanent settlements. They spread themselves thin over an enormous area, yielding to the danger of too great expansion. Their hold upon the country was akin to that of the small, widely scattered Indian tribes, whose manner of life they came to imitate and whose economic methods they adopted. Superficial use of a country's resources always means sparse population and therefore a weak hold upon the land. The compact settlements of the English made a surer basis for military operations than did the scattered posts of the French. The English had no mighty waterway to tempt them into the remote lands of the abundant furs. The peltries came to them on the upper Hudson and the Mohawk; but everywhere else the fertile river valleys of the narrow Atlantic plain rooted them in the soil, or numerous harbors and neighboring fishing-grounds bound them to the coast. Their advance towards the interior was therefore a slow heel-and-toe process, and never lost its connection with the tide-swept shore.

As the French settlements in North America were dominated by two great rivers, the English were controlled by the larger and smaller indentations of the coast, together with the streams draining into them. Coming from their island home with the object of making permanent settlements, they met geographical conditions which combined to strengthen them in the purpose. The need of keeping in touch with the mother country, at first for supplies and later for commerce, and the equal need of getting beyond the sandy beaches and rocky shores of the coast to the more fertile soil as well as to the trade of the interior, were the two motives determining the distribution of their colonies. Under the influence of the first, their settlements were grouped in inclosed sea-basins which carried their vessels by protected waterways as far as possible into the land. Massachusetts Bay, Narragansett Bay, Long Island Sound,

New York Bay with Raritan Bay, Delaware Bay with
the Delaware estuary, Chesapeake Bay, Albemarle with
Pamlico Sound, marked each the beginning of one, in
some cases two, of the original colonies.[7] The same
principle of selection was shown in the geographical
distribution of the French settlements in Acadia, where
the Bay of Fundy and Penobscot Bay comprised the
majority of their fishing-stations and trading-posts, and
their three agricultural villages of Port Royal, Beau-
bassin, and Les Mines.

All of these inlets answered the requirement of contact with the sea; the rivers flowing into them answered
the other requirement of broad contact with the land.
Some of these bays, like Massachusetts and Narragansett, which received no affluents, gave to their settlements
preëminently the stamp of coast communities. In such
the advance of the population towards the interior was
very slow, especially as there were almost no fertile alluvial bottoms to vary the monotony of a stubborn glacial
soil and to tempt to agricultural expansion. But wherever navigable rivers flowed into these inlets of the
sea, we notice the predominance of settlement at or
near their mouths and a steady advance of population
up their valleys. The mouth, and also in general the
lower course of a river, afforded safe harborage for
the small craft of the colonial period, while the back
country was made accessible by the main stream and
the ramifications of its creeks for canoe travel. A settlement thus located on a river commanded the Indian
trade of its whole basin, and could easily export the
products of its fields and forests. The instructions of

the London Company to the officers of their first expedition, sent to Virginia in 1606, point out another advantage of a river location well up towards the head of sea navigation : "Such a place you may perchance find one hundred miles from the river's mouth, and the further up the better, for if you sit down near the entrance, except it be in some island that is strong by nature, an enemy that may approach you on even ground may easily pull you out."[8] This caution was dictated by fear of coast attack from rival colonies of other European nations. The destruction of Ribault's colony at the mouth of the St. John's River in Florida by the Spanish, and later French depredations upon English settlements on the Maine coast, showed the wisdom of this precaution.

The river-made peninsulas of Maryland, Virginia, and North Carolina formed a geographical environment which influenced profoundly the development of these colonies. Here a mild climate, abundant alluvial bottoms, navigable streams, and a network of creeks combined to further agricultural expansion. Twelve years after the founding of Jamestown twenty-five miles from the mouth of the James River, the plantations extended up that watercourse for seventy miles, spreading out four or six miles from either bank. In a few years more (1624) population had pushed up the stream to the present site of Richmond and across the peninsula between the James and the York. The intervening waterways and the series of rivers made it easy to expand laterally, so that by 1663 we find the Virginians had spread northward to meet the Maryland colony on the Potomac, and southward to the Chowan Peninsula on Albemarle Sound. The ramifying streams brought almost all the plantations of this tidewater country in contact with the sea. It was customary for every planter to have his own wharf where he shipped tobacco or corn in exchange for the merchandise of Europe or the salted codfish of New England. Plantations on the upper reaches of small affluents were always able to send their produce downstream in canoes to the nearest head of navigation. Canoe and shallop became the coach of state and the vehicle for neighborly visits, while every plantation was its own seaport.

The result of these geographical conditions was a wide and evenly distributed rural population, lack of roads, and absence of large commercial towns. Nevertheless, certain natural features lying beyond the limits of Chesapeake Bay caused the rise of a bustling city at each extremity of this basin. The head of Chesapeake Bay just failed to touch the territory of Pennsylvania, much to William Penn's disappointment, for his colony had no coast and its one seaport town, Philadelphia, was quite on the eastern border. Hence much flour, farm produce, and quantities of valuable furs from the interior of Pennsylvania found their way down the Susquehanna River to Baltimore for export. This town, founded in 1729, grew so rapidly that in 1770 its population numbered 20,000. At the other end of the bay Norfolk came to be a fairly active seaport. Two different geographic factors were operating here. The town became an outlet for the produce of the Albemarle Sound country, whose rivers had their navigation spoiled by the prevalent sand-bars, and whose fine lumber from the pine barrens met a steady demand in Virginia, as the tidewater country was stripped of its forests. Furthermore, the location of Norfolk at the mouth of Chesapeake Bay gave it a certain commercial advantage and attracted the West Indies trade in sugar, molasses, and rum, which were exchanged for pork and beef of the Sound country, and subsequently marketed in the river plantations.[9]

Thus the early development of the French and English colonies in America was profoundly influenced by the watercourses. The New England settlements, committed by their geographical conditions to fisheries and a coastwise trade, found great compensation for their lack of rivers in the protected waterway of Long Island Sound. Their geographical location at the northern end of the English colonies, and the inevitable encroachment of their fishermen and traders upon the shores of Acadia brought them into conflict with the French and made them the natural bulwark against French aggression in this locality, just as the Hudson and the Mohawk in the middle section lent this same character to New York.

NOTES TO CHAPTER II

1. Ratzel, Zur Kustenentwickelung Jahresber. d. Geograph. Gesellschaft in Munich, 1894.
2. Ratzel, Anthropogeographie, part i. p. 248.
3. Weeden, Economic and Social History of New England, vol. i. 147.
4. Parkman, The Old Régime in Canada, p. 297.
5. La Salle's early discovery of the Ohio by way of the Allegheny River, even if authoritatively proved, would not invalidate this general statement, because sporadic and not productive of permanent results.
6. Justin Winsor, The Mississippi Basin, p. 22.
7. Ratzel, Zur Kustenentwickelung.
8. Quoted in Fiske, Old Virginia and her Neighbours, vol. i. p. 72.
9. Ibid. vol. ii. p. 211.

CHAPTER III

THE INFLUENCE OF THE APPALACHIAN BARRIER
UPON COLONIAL HISTORY

HISTORY shows us by repeated instances that the geographical conditions most favorable for the early development of a people are such as secure to it a certain amount of isolation. For this reason, a highly articulated continent like Europe has proved a forcing-house for nations. Almost every people there has grown up, shut off from its neighbors by barriers of mountain or sea. Confined to a limited area, protected from without by bulwarks of nature's own making, population increased rapidly and civilization moved with strides under the strongly interactive life. The people soon filled out their natural territory, then began to crowd it, pressing upon the limits of subsistence, and perfecting their political and social organization in the effort to avoid the friction incident to greater density of population. Local life grew in intensity and the sense of statehood was early developed. An increasing industrial and commercial activity endeavored to supply from abroad the deficit of food for the growing number of consumers; while the population, already redundant, began to expand beyond its natural environment and overflow into other lands.[1]

Of the three leading colonizing nations which came to

North America from Europe, one appropriated the only part of this continent which could afford geographic isolation in any way approximating that which it had enjoyed in Europe. That people were the English. At the end of the first century of permanent settlement they found themselves in possession of a narrow strip of coast, shut off from the interior of the country by an almost unbroken mountain wall. Sea and watershed drew their boundary lines and constituted at the same time their frontier defenses. Only one border was really open, that to the south along the Spanish possessions in Florida. The English were therefore in a naturally defined area, isolated enough to lend them the protection and cohesion which colonial life so much needs, affording the long line of coast which could give to this maritime people its most favorable environment, large enough for growth and strength, but small enough to secure concentration and to guard against the evils of excessive expansion. Beyond this seaboard country lay the great valley of the continent, shut in by the upheaved masses of the Appalachian and Rocky Mountain systems, a vast basin unbroken save by the faint traceries of its winding streams. Here nature offered no obstruction, afforded no protection. The two natural highways into this country, the St. Lawrence on the east and the Mississippi to the south, came into the hands of the French, and consequently gave them control of this extensive territory. It proved, however, too large for them to hold; the very extent of it scattered their population, tempted to the adventurous, half-nomadic occupation of the fur-trader rather than the

sedentary life of the colonist. Here were seen fortified trading-posts instead of the agricultural villages
which dotted the seaward slope of the Alleghenies.

The people who came to the New World to "plant
an English nation in America" had experienced at
home the economic evils of overcrowding. They realized that here was a chance to make a living; and
in so far their motive was wholesomely selfish. They
went to work accordingly to get an industrial hold on
American soil. The narrow strip of land between the
Atlantic and the mountains favored and strengthened
their purpose. The Appalachian barrier narrowed their
horizon and shut out the great beyond; it took away
the temptation to wide expansion which was defeating
the political aims of the Spanish and the French, and
transformed the hunter into the farmer, the gentle-
man adventurer into the tobacco-grower. Territory
that is held industrially in all its extent is held strongly.
The less dispersed the population, the fewer are the
avenues for invasion and the more solid is the front
which the country presents to attack. The mountain
wall gave to the Thirteen Colonies a certain solidarity
which they would not have otherwise possessed — a
solidarity which fought for them in the Revolution.

The Appalachian system, which, together with the
ice-worn highlands of New England, presented such an
insuperable barrier to the early colonists, extends from
the Green Mountains of Vermont to the pine-covered
hills of Alabama. It consists in general of parallel
ranges, altogether some three hundred miles in width,
which stretch along with only one considerable break

in all their length of thirteen hundred miles. A mantle of primeval forest with a singularly dense undergrowth contributed further to make them impassable. The backwoodsman had fairly to carve a path for himself through this wall of living green. In consequence, the tidewater country had its long-established colonies before anything of the mountains was known. The rivers flowing down the eastern slope were not navigable far back from the coast and therefore did not afford ready access into the interior; but when followed to their head waters they were found to disclose excellent passes. This was especially true of the southern portion of the system, but even here the disposition of the passes involved long, circuitous routes to reach the western slope; for, one range passed, the next one presented a similar barrier and the longitudinal valleys between had to be traversed before another gap could be found. The Pennsylvanians gained access to the Ohio by the West Branch of the Susquehanna, and also by another route further south from the Juniata to a tributary stream of the Allegheny. The Virginians, though, found a more direct way up the valley of the Potomac and thence by a short portage to the Youghiogeny. Among the broken hills of the southern end of the Alleghenies, an almost level route was frequented by the traders from the Carolinas and Virginia seeking the Cherokee villages; it was known from the earliest times, but had only a limited use, because it was too remote from the northwestern Indians who commanded the all-important fur trade.

The only important break in this mountain wall was

to be found in the natural depression of the Hudson and Mohawk valleys,[2] where the pass into the interior is only about four hundred and forty-five feet above sea-level. This route was also able to tap the northwestern fur trade, then in the hands of the French. Furthermore, trails led from the Mohawk and Genesee to the upper Allegheny, and thence to the Ohio and Mississippi. For this reason it became apparent at an early date that the Mohawk and Hudson valleys formed the key to the Northwest, as the meeting-place of the Allegheny and Monongahela was in reality the " Gateway of the West." By the geography of eastern America, therefore, these rivers were cut out for battle-grounds in all colonial wars between the English and the French, just as the head streams of the Po and Ebro leading down from the passes of the Alps and Pyrenees have been the scenes of conflict in every northern invasion of Italy and Spain since the days of Hannibal and Roland.

The French, who felt that the heart of the continent was in jeopardy, kept a sharp eye on these avenues to the West. They could attack the English most easily along the Mohawk valley and at the southern end of the mountains ; but at both these points the English had a buffer state between themselves and the enemy in the Iroquois tribes in the north and the Cherokees in the south, both of which nations were attached to British interests. The chronic jealousy between the French and English took an acute form when the French discovered that the Mohawk valley was getting too large a share of the northwestern fur trade. Consequently, by a long series of wars, they endeavored to drive the

English out of this region altogether. The enemy were at a long distance from the middle and southern colonies, which were further guarded by the wall of the Alleghenies; so that, with the exception of one or two sporadic attacks, they were left undisturbed by the struggle going on to the north of them. On the colonies along the Hudson and in New England, though, the attacks were almost incessant. In the French and Indian War nearly every foot of the upper Hudson was fought over as far as Lake Champlain, and the route up the Mohawk to Oswego was almost as bitterly contested. This history repeated itself in the Revolutionary War. Both in

MOHAWK VALLEY AND LAKE CHAMPLAIN

the French wars and in the Revolution the Six Nations who inhabited this region rendered valuable aid to the British. In the earlier wars particularly, the fact that they occupied a strategic position gave them a power and importance out of all proportion to their numbers.

The British early ingratiated themselves with the Iroquois and Cherokees, that they might have these

Indians as outposts against the French. But the rest of the tribes on the seaboard were not treated with any undue consideration, for they were not a numerous and therefore not a dangerous enemy; and they were prevented by the mountains from making any combination with the far more populous tribes of the Mississippi valley. The bodies of savages with whom the settlers had to contend were small. The Indians of southern New England were exterminated in two short wars. The Delawares were dislodged from their original home and emigrated beyond the mountains. The Tuscaroras were made to evacuate their holdings in North Carolina. Living in the narrow area of the tidewater country, it was unavoidable that the Indians should soon feel the encroachments of the whites; and it was equally certain that they would suffer defeat, when it came to a conflict, in consequence of the weakness of the tribes. Fortunately for the young colonies, the Alleghenies protected them against the depredations of the wild, half-nomadic Indians of the Northwest, and of the fierce Appalachian tribes of the southern Mississippi.

Hemmed in thus by the mountains, for the first one hundred and fifty years of their occupancy the English settlers were limited to the tidewater region of the Atlantic coast. This seaboard country presented in its different portions different aspects, which had a corresponding effect upon the colonists. In New England the lowland belt is only from fifty to eighty miles wide; but it gradually broadens as it continues southward, till in the Carolinas the mountains are two hundred and fifty miles back from the sea. The area adapted to settlement was therefore more extensive in the South than in the North. Furthermore, the northern district had suffered glaciation ; it was covered with a heavy deposit of boulders, which had to be removed at the cost of infinite labor before the land was ready for cultivation. The hardest work had to be done before a plough could be used ; but this once over, the soil could be tilled for a long time without giving signs of exhaustion.[3] This fact, together with the small area at their command, preserved to the settlers the contracted territorial ideas which they had brought with them from the mother country, and served to root them in the soil. Consequently the people of New England developed little of the tendency to expansion which later became a characteristic of the American people.

That tendency developed further south. Here the larger unobstructed area invited it, and the leading occupation of the settlers — tobacco-culture — made it a necessity. At a time when artificial fertilizing was almost unknown, the production of the better kinds of tobacco demanded a virgin soil. The planters were therefore led to take up as large tracts as possible. The only preparation was " girdling " the trees, the primitive mode of clearing the land which the colonists learned from the Indians. Land so prepared was planted in tobacco for three years, and afterwards in corn. This method of cultivation, expedient in view of the abundance of arable land,[4] was superficial, and the materials taken from the soil were never replaced. By this system of agriculture, the evils of which were further accentuated by slave labor, low lands were exhausted

in eight years, fields less favorably situated in three.
Ceasing to yield, they were abandoned and allowed to
revert again to a state of nature. More forest was
cleared for the plough, and the settlements invaded the
wilderness. In 1685, "although the population of
Virginia did not exceed the number of inhabitants in
the single parish of Stepney, London, nevertheless they
had acquired ownership in plantations that spread over
the same area as England itself." [5] Thus developed that
spirit of expansion which early in the eighteenth century led the settlers to hammer at the gates of the
mountain wall on their western frontier, and to resent
the claims of the French that the British possessions
were limited by the crest of the Alleghenies.

It was therefore not a matter of chance that the first
protest against the French forts on the Ohio was made
by a governor of Virginia. Towards the gap in the
ridge of the Alleghenies between the head waters of the
Youghiogeny and the Potomac, French and English
were approaching up the flanking valleys. Fort Cumberland was Maryland's frontier defense on the upper
Potomac to guard this strategic point; Fort Necessity
and the battle of Great Meadows represented the effort
of the Virginians to secure the western approach to the
dividing range, just as Braddock's march by this same
route a year later was aimed at Fort Duquesne, which
commanded all the river approaches from the west to
the mountain rampart of Pennsylvania, Maryland, and
northern Virginia.

The Virginians in the battle of Great Meadows took
the initiative in the French and Indian War before the

declaration of hostilities.   Men like Byrd of Westover
and Governor Spotswood appreciated the character of
the mountains as a bulwark against the enemy, and realized the necessity of making themselves masters of its
passes before the French should do so.   They knew
that their expansion over these mountains was inevitable; therefore the sea-to-sea claims of the English were
to them of vital importance.   This view was shared also
by the colonists of Pennsylvania and New York because
they, too, lived under similar conditions of expansion;
but the people of New England, on the other hand,
were indifferent to the disposition of the western country.   They seemed quite satisfied with the line of the
Alleghenies as a boundary, if the trade with the western
Indians could be secured.   Their standpoint was therefore provincial, in contrast to the continental concep-
tions of the middle and southern colonies.

The policy of the latter prevailed and the whole line
of settlements from Maine to South Carolina was levied
on for its support in the French and Indian War.   This
was the first time in their history that all the colonies
acted together; this was the first time that there was
a common American interest at stake.   Hitherto none
had combined, except the New England colonies, where
the geographical conditions made for greater density
of population, and a certain degree of isolation emphasized their community of interests.   The southern colonies, with their widely scattered plantations, did not at
first have much to do with each other or with their
northern neighbors.   Blocked by the mountains in their
growth towards the west, however, they were finally

compelled to expand laterally and fill up the stretches of forest which originally separated them. The effect of the Appalachian barrier was therefore to keep the population within clearly defined limits and lend it that density which means strength. In 1700 " it was possible to ride from Portland, Maine, to southern Virginia, sleeping each night at some considerable village." [6]

With the rapid increase of population characteristic of colonial life, the limited extent of the country which would repay cultivation, and the prevailing wasteful methods of agriculture, it was to be expected that the supply of arable land would eventually be exhausted, and the activities of the colonist necessarily directed into other channels. This condition was reached first in New England, so that its people were early forced into industrial and maritime enterprise. Manufactories for the most common articles of consumption were established ; the forests were levied on for shipbuilding, and American vessels were soon doing all the carrying trade between the colonies.

The Appalachian barrier had the effect, therefore, of keeping the colonies to the hem of the continent. It limited them to a strip of coast, where they were most easily retained under British domination. If they had expanded at an earlier date to the west, England would have found it a much more difficult task to make her power felt over all the area settled. The mere element of distance in a new, unbroken country would have complicated greatly the machinery of government, while diminishing its efficiency. Furthermore, it would have generated in a larger proportion of the population the spirit of the frontier, that is, the spirit of independence. Disaffection towards the mother country would have developed slowly as a chronic disease; and we may think it would have been a long time before the feeling could have gathered strength to break out in rebellion. As it was, held under the thumb of the British government, disaffection took on the form of an acute attack, and rapidly ran its course from protest to rebellion, from rebellion to independence. More than this, when the conflict did come the colonies were all of one mind, no matter from what section they came. The spirit of union which animated them can be attributed in no small degree to their close contiguity, while their occupation of a contracted area with their two and a half million population enabled them to operate in a solid mass against the enemy.

In revolting against England the American colonies followed a recognized law of political geography. They constituted the remote western frontier of Europe; and a tendency towards defection manifests itself in all peripheral holdings.[7] History is full of examples. The causes are deep-seated. Differences of geographical conditions, of climate, soil, economic methods, and therefore of political and social ideas, rapidly differentiate colonists from the parent nation. Moreover, mere distance increases greatly the difficulty of governmental control, even in this day of rapid communication, as England has experienced recently in Cape Colony. A hundred years ago Burke stated this politico-geographical law in terms that cannot be improved upon. " The last cause of this disobedient spirit in the colonies . . .

is deep laid in the natural constitution of things. Three thousand miles of ocean lie between you and them. No contrivance can prevent the effect of this distance in weakening government. Seas roll and months pass between the order and the execution; and the want of a speedy explanation of a single point is enough to defeat a whole system. . . . Nothing more happens to you than does to all nations who have extensive empire; and it happens in all the forms in which empire can be thrown. In large bodies, the circulation of power must be less vigorous at the extremities. Nature has said it. The Turk cannot govern Egypt and Arabia and Kurdistan as he governs Thrace; nor has he the same dominion in Crimea and Algiers which he has at Brusa and Smyrna." [8]

This is the statement of the law from the standpoint of the governing power; those on the far-away periphery also were affected by their remoteness from the center of authority. The colonists found it difficult to get a hearing in England; distance dulled the edge and weakened the force of their protests, — this by a psychological law. They found it irksome, often detrimental to their interests, to wait months for the ratification in England of colonial laws. Compelled often by sudden crises to act without authority in consequence of their remoteness, the colonists developed a spirit of initiative and independence. When the outbreak came, England for the first time learned the expense and difficulty of an arm's-length war. Time and space fought on the side of the Americans.

With the opening of hostilities, geographical conditions directed campaigns and influenced the result of battles. The valley of the Hudson made the natural line of communication between Canada and the British fleet on the coast, and at the same time formed the land connection between New England and the other colonies. As the British had full control of the sea, the colonial armies could move only by land routes. Hence the Hudson River was of vital importance to both sides, and from the first both sides aimed at its control. The seizure of Crown Point and Ticonderoga at the start enabled the Americans to command the line of communication with Canada. The British operations to gain the Hudson followed three geographical lines. By the Champlain route and the Mohawk valley two armies from Canada advanced towards Albany, where they were to form a conjunction with troops coming up the Hudson. The naval battle off Valcour Island in Lake Champlain, and later the capture of Ticonderoga and Fort Edward, traced the British advance to the Hudson from the north; Bemis Heights, Stillwater, and Saratoga, the attempted retreat along the same route. On the Mohawk valley route, Oswego, Fort Stanwix, — which corresponded in its geographico-strategic location to Fort Cumberland on the upper Potomac, — and Oriskany marked the conflict between the British and American forces. The approach from the south began with the operations at the mouth of the Hudson. Here the siege of Brooklyn Heights, battles of Harlem Heights and White Plains, with the capture of Fort Washington, gave the British the control of Manhattan and the lower Hudson. But the success of the

Americans in cutting the British lines of communication with Canada along the Mohawk and Lake Champlain, and Washington's tactics to prevent Howe's moving up the Hudson, frustrated British effort to control eastern New York.

The colonies by their position on the long, narrow border of the American continent had an extensive sea frontier, which was open to the attack of the British fleet. Consequently much of the Revolutionary War was fought on or near the coast. The coast towns of New England were cannonaded and levied on for supplies by the British frigates. Charleston, South Carolina, and Savannah were taken by naval attack, and the fleet assisted in the capture of New York. With the exception of Canada, the British had no base on the land frontier, so they had to establish their base at coast points accessible to their ships. And if their armies advanced into the interior, a threat to their line of communication with the sea would bring them hurrying to the coast. Washington's occupation of Morristown Heights in January, 1776, enabled him to prevent the enemy from crossing New Jersey to take Philadelphia, because he was in a position to cut them off from their supplies. The campaigns of Cornwallis and Tarleton in the Piedmont country of North and South Carolina were terminated chiefly by the threat to their lines of communication with the coast. The final overthrow came when, by the advent of the French fleet, the Americans were able to invest the position of Cornwallis at Yorktown by water as well as by land.

The barrier of the Appalachians kept the American

armies facing eastward. The colonies braced themselves against the mountains and fought towards the sea. The bulwark at their back protected them from the onslaughts of the western Indians, who were stirred to hostilities by British agents. Only the few settlements beyond the mountains in Kentucky and Tennessee were exposed to this danger; but as they were debarred in general from participation in the eastern campaigns, they could give the Indians their undivided attention.

(For a hundred and fifty years the American people were dammed up against the mountain barrier. But the energies aroused by the prosecution of a successful war, and the snapping of the cords which held the colonies in leash to England, enabled the mass of American life to rush through the breaches in the mountains, down to the Mississippi and beyond; till in half the time it had taken the people to reach the crest of the Alleghenies, they were planting their towns on the genial coast of the Pacific.)

NOTES TO CHAPTER III

1. Shaler, Nature and Man in America, p. 152, and Ratzel, Anthropogeographie, part i. p. 249. Second edition, 1898.
2. A. P. Brigham, The Eastern Gateway of the United States, in the Geographical Journal, May, 1899, reprinted in The Journal of School Geography, April, 1900.
3. Shaler, Nature and Man in America, p. 225.
4. Roscher, Nationalökonomik des Ackerbaues, pp. 74 and 104. 12th edition.
5. Philip Bruce, Economic History of Virginia, vol. i. p. 424.
6. Shaler, Nature and Man in America, p. 199.
7. Ratzel, Anthropogeographie, p. 219. 1899.
8. Burke, Speech on Conciliation, Bohn Library, vol. i. p. 468.

CHAPTER IV

A MOUNTAIN system is always a barrier; this character
it maintains in a greater or less degree according to its
physiographic features. A single high massive range
like the Himalayas, or a double snow-crowned rampart
like the Great Caucasus and Anti-Caucasus, or a series
of parallel earth-folds like the Alps, are enduring obstacles to intercourse, whose resistance is measured
largely by their height and width. The number, distribution, and elevation of its passes also affect the
barrier nature of a mountain. A comparatively low
range like the Pyrenees, which lifts an even sky-line
unbroken by any deep gap or notch, is a wall without
breach or door, while the many easy and evenly distributed passes of the Alps have reduced the difficulty
of communication between the plains to the north and
south. Finally, the presence of transverse river valleys
leading up like natural roads to these lofty portals in
the mountain wall facilitate transmontane travel, and
their absence leaves many an excellent pass unutilized.

The general with his army, the pack-laden trader,
migrating settler, and the engineer laying out the course
of a railroad, see in a mountain range only a challenge
of their energy and endurance; and they straightway

search for the lowest dip in the crest by which to cross. The pass points the easiest road across the ridge, and therefore becomes the focus of all established highways on either slope. Along its narrow, rugged channel pours a tide of humanity, drawn thither by some dream of good beyond, while on the flanking summits reigns the solitude of primeval forest or the eternal snows.[1]

Mountains which spread out on a broad base with a succession of parallel ranges must be crossed by a succession of passes determining more or less circuitous routes. This is the case in the Austrian Alps and in the Hindu Kush along the caravan road from Peshawar in the Punjab to Balkh in the plains of northern Afghanistan. But sometimes powerful drainage streams have carved their valleys straight across the axis of the corrugated upland and opened up natural highways to a single pass, which thus dominates a system several hundred miles wide. The valley of the Ticino from Lake Maggiore and of the Reuss from Lake Lucerne form an unbroken ascent to the St. Gotthard, the only point in which the central Alps can be crossed by one pass. But even where streams have not cut reëntrant valleys into the heart of the mountain system, but furrow only the outermost slope, they mark the avenue of easiest approach to the gaps above. Navigable in their lower course, they expedite travel; as the grade grows steeper, however, rapids demand portages, and later canals, but always the rugged gorge above furnishes a foothold for the trail to the summit.

By river valley and pass migrating races and individuals have found their way over mountain barriers.

The Gap of Belfort, connecting the long valleys of the
Rhone and the Rhine, has been the historic route of
migration and travel between the North Sea and the
Mediterranean. The Brenner Pass was the Alpine
portal by which the Cimbri swept down into the valley
of the Po ; and the breaches in the eastern Alps determined the line of Rome's expansion to the Danube, just
as later they exposed her to the inroads of the Goths and
Huns. To-day the Russians are pushing southward
from the sandy plains of Turkestan up the Heri Rud
and Murghab rivers to the low dips in the western
Hindu Kush which form the " Gates of Herat ; " and
eastward by the valleys of the Syr Daria and the Oxus
she has stretched her power to the snowy passes of the
Tyan Shan and the Pamir, from which she hopes to lay
hold on Mongolia and Chinese Turkestan. In the same
way, gaps and rivers enabled the American colonists to
cross the Appalachian barrier. The location and distribution of these natural passways determined which colonies should furnish the largest quota of the pioneers,
and also what should be the destination of those early
winners of the West. To see how this geographic control operated, an analysis of the Appalachian topography
is necessary.

The Appalachian system is peculiar in that it consists
of a long central zone of depression, bordered on the
southeast by the Appalachian Mountains proper, and
on the northwest by the Allegheny and Cumberland
plateaus. This central trough, known as the Great
Appalachian Valley, is sunk several hundred feet below
the highlands on either side, but its surface is diversi-

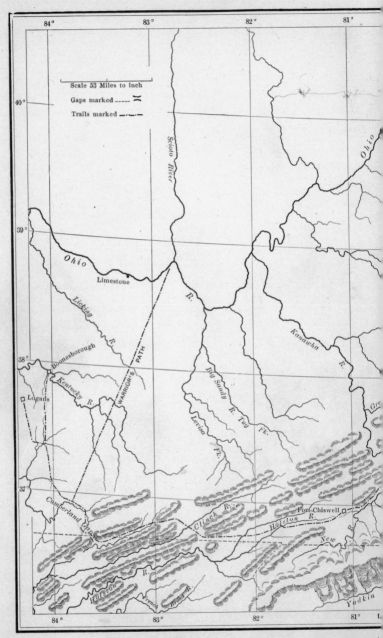

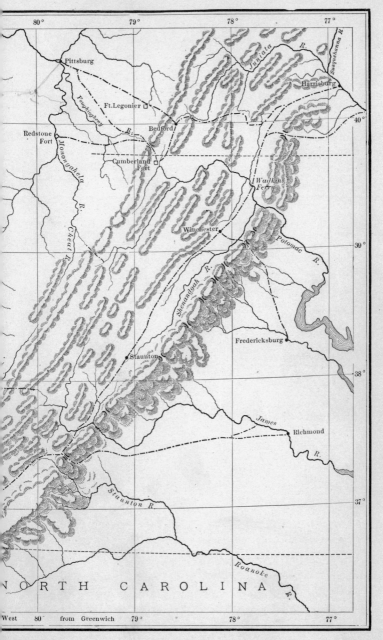

fied by intermittent series of parallel, even-crested ridges, which rise, one beyond the other, like successive waves of the sea, across the whole seventy-five miles of the valley's width. The Blue Ridge, which forms the eastern rim of the Great Valley, is a range of gentle slope and rounded crest, notched at frequent intervals by gaps a thousand feet deep, or cut through by flaring river gorges. The western rim is the Allegheny Front, the bold, forbidding escarpment of the Allegheny and Cumberland plateaus, whose rugged surface, scoured and furrowed by drainage streams, runs off into the low plains of the prairies and the Great Lakes. (The height of the Appalachian system increases from the north towards the south ; but the average elevation of the ranges, which is only from three to four thousand feet, is nowhere sufficient to constitute a great obstacle.) It is the long, unbroken extent of the system and its great width of three hundred miles which make it a barrier.

This character would be far more pronounced were it not for the drainage streams of the upland. The Appalachian rivers cut across the grain of the mountains from west to east or east to west, only intermittently controlled by the ranges and following the trend of the longitudinal valleys. The reason for this physiographic discordance is to be found in the geologic history of the highlands.[2] The transverse river channels are older than the mountains themselves, and as the nascent ranges slowly protruded their crests, or were left in relief by erosion, cut through them in flaring water-gaps. The northern rivers, like the Delaware, Susquehanna, Potomac, and the James, have their sources on the

Allegheny Plateau, cut through the Allegheny.Front, the ranges of the Great Valley, and the Blue Ridge or its northern extension, South Mountain, and empty into the Atlantic Ocean, while their head waters interlock with those of the Ohio along inconspicuous divides to the west. The southern rivers take an opposite course. They rise, as the Watauga, Nolichucky, Holston, French Broad, and New River, on the eastern rim of the system in North Carolina or, as in the case of the New, even on the seaward side of the Blue Ridge, carve their way westward through every barrier, and flow out as the Tennessee and Kanawha into the Ohio. Their sources interlace with those of the Yadkin and Catawba, which furrow the eastern slopes of the Blue Ridge to the Atlantic; and among the Allegheny ridges of West Virginia, the Greenbrier, a northern tributary of the Kanawha, is separated from the head waters of the James River only by a narrow range tracing the Allegheny Front, which is now crossed by the Chesapeake and Ohio Railroad.

The divides within the Great Valley are peculiar in being distinct from the Allegheny ridges; for this reason they are intricate and little elevated above the general level, so that it is easy to pass down the whole length of this trough from the Susquehanna to the Tennessee, merely by following the lateral branches of these and of the intervening streams as they flow along the minor longitudinal vales of the whole system. Hence the outbound pathfinders to the west were able to enter the Great Valley by the Susquehanna, Juniata, Potomac, James, or Roanoke, turn right or left up a lateral branch ascend a notched gap or low watershed to the next transverse river within the valley, and following it to its source pass over to some westward flowing stream. In this way he could cross the Appalachians in almost any direction.

The eastern head waters of the Tennessee, after leaving the high open valleys just to the west of the Blue Ridge, had to cut through the broad, complex belt of the Unaka Mountains [3] in deep gorges, which afforded no easy passway and were further dangerous on account of the hostile Cherokees. Hence North Carolina pioneers from the upper Watauga and Nolichucky valleys found readier connection with the west through the New River and the Virginia routes to the northern tributaries of the Tennessee. And for over a hundred years the Unakas remained an obstacle to deflect the tide of emigration.

The westward progress of the colonial frontier in early times was slow. Bound to the coast by the need of ready sea communication, in 1700 the settlements embraced all the tidewater country up to the "fall line," the outer limit of the Piedmont Plateau and the first interruption to river navigation; the frontier was somewhat farther inland, about fifty miles to the east of the Blue Ridge. But by 1750 the far-ranging traders were building their lonely cabins in the murmurous solitude of the great Appalachian forests, while their pack-horses were loosening the rocks on the mountain passes beyond, and their bateaux were burdening the waters of the Ohio, Great Miami, Wabash, and Sandusky Bay.[4]

The successful ending of the French and Indian War,

which extended England's territory in America to the Mississippi (1763), was the signal for a western advance of the population; but at the outbreak of hostilities (1755) the frontier of settlement described a rough curve, which reached its easternmost points in New York and Georgia, and its westernmost on the Greenbrier and Holston rivers. Conditions of physical and political geography determined this frontier line. In New York the advance up the Mohawk valley moved at a snail's pace as compared with the leap of settlement from Manhattan Island to Albany. From Schenectady some late-coming Germans pushed the narrow valley zone of civilization constituting New York first to Palatine Bridge and Stone Arabia (near Canajoharie), then in 1723 to German Flats along the upper Mohawk between Little Falls and Utica. The Catskill and Adirondack mountains on either side, too inaccessible in those roadless days, were left uninhabited. From 1723 till 1784 there was no advance of settlement in central New York, for, though the country was fertile, the location was too exposed to French aggression; and after the removal of this danger, the relatively dense population of the Iroquois tribes was an effective barrier to expansion. Thus political neighbors held back the frontier of settlement.

In eastern Pennsylvania, where the Appalachian trough is easily accessible in consequence of dips and breaks in the mountain wall, settlers early spread into the Cumberland and Lebanon valleys, and founded Easton, Bethlehem, Reading, Carlisle, and Shippensburg, while the traders had pushed up the Susquehanna to Wyoming and across the " Endless Mountains " by the Juniata, in whose upland valleys they were marking their "tomahawk claims" and preparing for permanent settlement.[5] But this western part of the Great Appalachian Valley is much broken by linear ridges and little adapted for agriculture. Hence the tide of frontier settlers, who were made up chiefly of Scotch-Irish and German late-comers to the New World, spread southward along the line of least resistance into the broad open Valley of Virginia. A few passed thence by the gap of the Roanoke River to the head streams of the Yadkin in the Piedmont of North Carolina.

Both these regions received accessions also from the tidewater portion of the southern colonies. Especially in Virginia, as the large estates predominated more and more, the small proprietors, unable to compete with the plantation system of the tidewater, moved westward and southwestward into the Piedmont and mountain regions, where other economic conditions prevailed, and thus by natural selection added to the democratic element of the frontier. The broad, fertile Shenandoah Valley was easily accessible to these through the low passes of the Blue Ridge between the Potomac and the Roanoke, and hence drew steadily if not extensively from the tidewater sources; but it received the greater part of its population by the nature-made highway leading down from the north. The geographical order of its settlement indicates this. The first clearing was made in 1732 near the present town of Winchester, about thirty-two miles from the Potomac, by one Joist Hite, who came in from Pennsylvania and who was soon followed by sixteen families from the same state. Two years later some cabins and cornfields appeared in the vast encasing forest about twenty-eight miles further up the valley in the vicinity of the modern Woodstock; and from then till 1740 the tide advanced rapidly up to the sources of the Shenandoah River, and beyond to where the head waters of the James and Roanoke interlace with those of the westward flowing New River.

In 1755, just before the outbreak of the French and Indian War, Winchester, Staunton or Augusta Court House, as it was then called, and Fincastle were distant frontier posts; but fifty miles beyond, the westward flowing streams, readily accessible from Virginia, showed the remotest habitations of all the colonies.   Colonel Wilson's mill on Cheat River was reached by a twenty-mile trace from the South Branch of the Potomac.   Just to the southwest there were a few cabins and clearings on the Greenbrier.[6]   William English or Ingles, with a few other Scotch-Irish and a group of Dunkards, had their outlying dwellings on the upper New far up the Valley of Virginia; and just across the divide to the south the enterprising Stalnaker — his name tells his origin — was building his cabin on a head stream of the Holston near the present Tennessee boundary, when Dr. Walker passed that way in 1750 to explore the lands of Kentucky.   In North Carolina at this time the remotest settlements were along the sources of the Yadkin; but just beyond were the westward flowing streams of the Tennessee, and here the Watauga and Nolichucky valleys in 1758 saw the first stockades across the Blue Ridge south of the Virginia line.[7]

So long as colonies were limited to the tidewater,

Piedmont, and the eastern part of the Great Valley in Pennsylvania, the Appalachians were an effective barrier against the western Indians ; but when the enterprising settlers began pushing out to the inner ranges of the mountains and contesting the territorial claims of the French beyond, the latter stirred up the savages to depredations upon the outlying settlements. Then every stream which opened a route into the mountains from the west became an Indian war-path ; and between 1754 and 1759 every mountain pass or divide at their sources began to bristle with English fort or stockade to check these onslaughts. In this period a whole line of such strongholds was erected, — Ligonier on the Loyalhanna branch of the Allegheny near the pass from the Juniata, Fort Cumberland on the upper Potomac, Fort Chissel in the Valley of Virginia on the divide between the New River and the Holston, Colonel Byrd's fort at Long Island on the upper Holston in East Tennessee, and a hundred miles further south Fort Loudoun at the junction of the Jellico and Watauga rivers. These remote military outposts and the scattered clearings about them were from a hundred to a hundred and fifty miles west of the frontier of continuous settlement in Virginia and North Carolina.

In the tidewater and Piedmont regions, the general movement of the population was northwest along the course of the drainage streams, at right angles to the coast ; but, within the mountains, population advanced down the longitudinal valleys, along a northeast-south-west line, parallel with the trend of the Appalachian system, and it swept along in its tide all the little tributary streams of outbound settlers who crossed the Blue Ridge. The consequence was that while the tidewater regions of the colonies kept each its distinctive character, born of distinct old-world sources and diverse environments, the backwoods population of the mountains, from the Wyoming valley to the Yadkin, showed a wide mingling of ethnic elements — Dutch, German, Huguenot French, Scotch-Irish, and English, — which obliterated the distinctive types of the coast, while the prevailing similarity of their geographic environment operated to produce the new type of the backwoods. From northern Pennsylvania to Georgia, the whole Piedmont and mountain tract gave its stamp to the life of this forest frontier. Mountain economy, based upon the small farm, did not permit the development of large estates, with the concomitant industrial system of slavery and aristocratic organization of society. Difficulty of transportation to the coast prevented trade on a large scale and the amassing of wealth. Without social classes and wealth, democracy reigned supreme. The daily struggle for existence amid the dangers of the wilderness produced a race of men, sturdy in their self-reliance, self-respecting in their independence, quick to think, strong to act, and above all filled with the spirit of enterprise. Their remoteness from the arm of the law led them to frame laws for themselves, or to take the law into their own hands. Therefore with impartiality they were prone to deal summarily with horse-thieves and tax-gatherers. But their hardihood and fertility of resource, practical and political, fitted them to be pioneers and founders of states in the more distant wilderness beyond.

When the westward movement came, the base of its advance was geographically defined. The isolation of New England excluded her from it. New York also took no share in it because of her French and Iroquois neighbors. Moreover, after the war with France, England was bound to regard the rights of her Indian allies. Therefore, by the treaty of Fort Stanwix in 1768, though the Iroquois ceded the western lands lying between the Tennessee and the Ohio, from Kittanning on the Allegheny the " Property Line " defining the frontier of white settlement turned eastward along the West Branch of the Susquehanna, crossed by a small creek to the North Branch, then to the elbow of the Delaware at the present site of Deposit, from which it turned directly north along the Unadilla branch of the Susquehanna to a point midway between Fort Stanwix and Lake Oneida. Western New York was thus secured to the Iroquois ; but even after the Revolution, which had made the Six Nations the enemy of the colonists, the retention of the Lake posts till 1796 by the British and the strength and number of the Indians themselves deterred the advance of settlements, which in 1800 scarcely extended to the Genesee River.

Around the southern end of the Appalachian Mountains an almost level route led to the Mississippi Valley, but here, as in the north, the Indians barred all western advance. Creeks and Cherokees were stirred by the Spanish, who acquired Louisiana in 1762, to drive the colonists in this region back to the coast. Later, crowded to the south and east by the encroachments of the whites in Tennessee and the infant state of

Alabama, these Indians were able to limit Georgia's frontier of settlement to the Altamaha and Ocmulgee rivers as late as 1820.

Pennsylvania, Maryland, Virginia, and North Carolina, therefore, were the only participants in the early westward movement. In the three southern colonies, land-hunger resulting from existing agricultural methods supplied the motive, while in Pennsylvania the far-ranging traders had drawn attention to the fertile regions beyond the mountains. Individuals from this colony and from Virginia were prominent in the land companies which from 1750 were trying to secure patents to large areas on the "Western Waters." [8] Furthermore, the provincial governments, with the approval of the British crown, issued innumerable warrants for land as pay for military service in the war with the French and with Pontiac ; such claims were to be located along the Ohio, within the charter limits of Virginia and Pennsylvania. Later, the young Republic, rich in land but poor in specie, paid off in the same way a part of the debt incurred in the Revolution, and thus gave a new impulse to the outmoving tide.

Geographical conditions favored the expansion of the middle colonies. Their long line of western frontier was a chord subtending the rude arc formed by the Tennessee and Ohio rivers. The head waters of these streams or of their tributaries opened so many gateways to the transmontane lands. The broad territory between the two rivers had been ceded to the English by the Indians at Fort Stanwix (1768). Reserved by the savages as a hunting-ground, and having, therefore, no permanent Indian settlements, it opened a line of least resistance for colonial expansion; while its fertile soil, agreeable climate, and abundant salt-springs afforded all essentials for pioneer homes.

The topography of the Appalachians and their two distinct drainage systems determined that the northern routes should converge where the Allegheny Plateau is spanned by the forks of the Ohio, and the southern where the Tennessee, Kanawha, and, just between the two, Cumberland Gap open portals to the West. Three routes met at Pittsburg : one from Philadelphia by the West Branch of the Susquehanna, a forty-mile portage over the divide, and Toby Creek to the Allegheny at Kittanning; a second further south, also from Philadelphia, by the Juniata tributary of the Susquehanna, or by a more direct trace known as Forbes's Road from Carlisle through Shippensburg, Fort Lyttleton, and Fort Bedford to the upper Juniata, thence by an easy mountain pass to the Loyalhanna River by Fort Ligonier and on down the Allegheny or across the low dividing ridge to the forks of the Ohio; and a third up the Potomac to Fort Cumberland and thence by Braddock's road over the divide to the Youghiogeny or to Redstone Old Fort on the Monongahela. This was the natural line of connection with Alexandria and Baltimore. On these northern routes water-carriage predominated. Navigation on both the Susquehanna and Potomac was interrupted by rapids; but for early canoe travel this obstacle was not serious. As the transmontane region became settled, however, the Potomac was improved by canalization and a better road shortened the distance between

Fort Cumberland and Redstone. The traders said that Alexandria was four hundred miles nearer the fur-fields than any other Atlantic port.[9] Therefore Baltimore received a large part of the bulky exports from the back country, while Philadelphia furnished to the western trade most of the manufactured supplies, which were sent by a more direct wagon-road three hundred miles across the mountains.

Once over the mountains, the pioneer continued his journey westward from Pittsburg by flatboat or keelboat down the Ohio. But as there was no established carrying business on the river at this time, the supply of even these primitive means of transportation was uncertain, all the more as, once at their downstream destination, they were unfit to stem the swift current on a return voyage. The emigrant, therefore, was often compelled to wait a month or more before he could get a boat built; for workmen were scarce. His troubles were not over when his rude craft was launched. Danger from Indian attack was constant and imminent. But after the Revolution and Wayne's defeat of the northern savages in 1794, and especially after the opening up of the Northwest, the Ohio route was much used, except for the return trip to the East. The current that bore the outbound settler down the tide of the western waters was his obstacle when he turned his face towards the coast. Hence this difficulty and the horror of Indian attack deflected much of the westward movement to the land route from southwestern Virginia by Cumberland Gap. The intervening trail from the head streams of the James River to the Greenbrier tributary of the Kanawha, and the rugged portal opened to the same river and to the Tug Fork of the Big Sandy by the deep cañon of the New, seem to have been little used by the pioneers. Buffalo trace and Indian war-path had pointed out the easiest way across the mountain barrier, and these the backwoodsman followed. The Big Sandy and New River route was the common war-path of the Shawnees when making incursions from Ohio into the territory of their relentless foes, the Catawbas of North Carolina, because it was the most direct line from the villages along the Scioto ; but the light equipment of savage warriors permitted rougher traveling than the pack-horses and cattle herds of the outbound settler.

The Cumberland Gap route was the natural avenue to the West for emigrants from Virginia and the Carolinas, but it was preferred also by colonists from Philadelphia when they carried little baggage, though the distance from that city to the interior of Kentucky was eight hundred miles. From Philadelphia an established line of travel led across the Potomac by Wadkin's Ferry, and up the Valley of Virginia along the old war-trail of the Iroquois and Cherokee, over the low watershed to the New River. The pioneer crossed that stream and continued up its western affluent, Reed Creek, which on an almost level divide interlocks with the head streams of the Holston. Here the western trail was joined by another path from Richmond, Virginia, and here at the " forks of the road " was Fort Chissel, the block-house built in 1758 to hold the Cherokees in check. At this point began the Wilderness Road. The distance to Cumberland Gap was two hundred miles.

From the upper Holston, the Wilderness Road turned west, and by a maze of gaps and their approaching streams which furrowed the mountain sides, it crossed the parallel ranges of Clinch, Powell, and Walden mountains to the Powell River, and turned down this valley to Cumberland Gap, an old "wind-gap" which opened an easy gateway (1600 feet elevation) through Cumberland Mountain to the West. Just beyond the pass the frontiersman struck the "Warriors Path," an Indian trail which ran between the Shawnee villages at the mouth of the Scioto in Ohio and the Cherokee lands in eastern Tennessee. The Wilderness Road, as tracked in 1775 by Daniel Boone for Colonel Henderson, followed this Indian trail across the ford of the Cumberland, where this river breaks through Pine Mountain, and down the stream for a few miles to Flat Lick; but here it turned northwest, and followed a buffalo trace along the ridges over to Rockcastle River. In Kentucky the pioneer, following the example of the buffalo, avoided the immediate watercourses; for in contrast to the broad basins of the Allegheny rivers, these streams had carved out the surface of the Cumberland Plateau into deep V-shaped valleys, which afforded only precarious foothold for the traveler and necessitated continuous crossing of their rushing currents.

By buffalo trace the road continued north from the Rockcastle River through Boone's Gap in the rugged barrier of the Big Hill range (present route of the Louisville and Nashville Railroad) to Otter Creek and the Kentucky River at Fort Boonesborough, thence to Lexington and the smiling lands of the Bluegrass.

From Rockcastle River another branch of the Wilderness Road, blazed by Logan in 1775, turned northwest and, by a natural gateway near Crab Orchard, reached level land near the present town of Stanford, where Logan built Station St. Asaphs. This track became more important than Boone's trail to the north because it led more directly to the attractive level lands of Kentucky, and, passing through Danville, Bardstown, and Bullitt's Lick, terminated at the Falls of the Ohio, whence was the readiest connection with the old French trading-posts on the Mississippi and the Wabash.

The route taken by Robertson to the Tennessee country followed the Wilderness Road beyond Cumberland Gap, and then turned southwest, guided largely by buffalo traces seeking pools and salt-licks, to the " bend of the Cumberland" where Nashville grew up; but the women, children, and baggage for the new settlement made a long and dangerous journey in flatboats, dugouts, and canoes down the winding course of the Tennessee River to the Ohio, and up the Cumberland to the little stockade on the bluffs. But later (1783) a new road from the confluence of the Holston and the Clinch rivers passed by easy ascent over Cumberland Mountain to the valley of the Cumberland and Nashville. This route was joined by a trail also from North Carolina, and at the mouth of the French Broad River by still another from South Carolina. Thus several roads from the east converged upon the upper Tennessee, just as Cumberland Gap was a plexus of other routes aiming at Kentucky and the West.

At first only blazed trails through the great wilder-

ness, then traces trodden out more plainly by the feet of
pioneers and pack-horses, impassable to wagons, these
transmontane roads offered no easy conditions for travel.
The floor of the Great Valley in southern Virginia rises
to 1700 feet and the linear ranges to be crossed were in
general only 600 or 800 feet higher; but each in turn
barred the western horizon like a wall, and only here and
there were their even sky-lines notched by a gap. These
gaps were never opposite one another in the successive
ranges, so the traveler had to take a circuitous way up
and down the intervening valleys from pass to pass. He
had to ford trough streams and mountain creeks, which
one summer rain might raise to rushing torrents. For
food he depended upon game or the cattle which he
drove along with him, while an outlying habitation occasionally supplied him with bacon and corn-meal. The
danger of attack from the Cherokees, stirred to hostility
first by the French and later by the Spanish, was always
there, but never so imminent as the perils on the Ohio
from the Shawnees. But in spite of dangers and hardships, the trail through the wilderness had its joys,—
the charm of the wondrous Appalachian forests, the
flicker of sunlight through the high-reaching trees, the
plunge into a tunnel of green through the tender spring
underbrush, the sense of strong, pulsing life with the
upward climb, finally the deep-drawn breath on the
summit before the outstretched billows of land, and
the hope of opportunity beyond.

The distribution of trans-Allegheny population in
1790 was in close relation with the western highways.
Pennsylvania settlements formed a continuous line from

the Juniata to the Allegheny, Monongahela, and Ohio
almost to the mouth of the Muskingum River. In
Virginia there was an unbroken area of settlement to
the western rim of the Great Appalachian Valley; but
beyond the Greenbrier and the New, the frontiersman

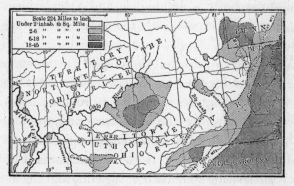

TRANS-ALLEGHENY SETTLEMENTS IN 1790

had passed by the rugged country of the Cumberland
Plateau along the upper Kanawha to make a small settlement at its mouth. The frontier farms and villages
of southwestern Virginia merged into "the independent
state of Franklin" in the parallel valleys of the Holston
and the Clinch. Downstream moved the tide of settlement. Where the Holston and French Broad unite to
form the Tennessee, White and Conner in 1787 located
a warrant of land which they had received as pay for
military service in the Revolution, and about their fort
the town of Knoxville was laid out soon afterwards. In
1785 a settlement was attempted at the Muscle Shoals
of the Tennessee River in northern Alabama, and a few
years later Sevier and others secured a grant of land for
settlement just south of the Shoals. The attraction here

was a geographical one.   The obstruction to navigation
by the rapids necessitated a carry at this point, which
therefore became a halting-place for travelers.   Furthermore, the Big Bear River, a southern tributary of
the Tennessee, connected by a short portage with the
head streams of the Tombigbee and Yazoo, thus opening up lines of trade with the lower Mississippi and
Mobile Bay.

To the west Nashville, whose eighty or ninety cabins
were a center for the five thousand pioneers settled for
eighty miles along the river, was a product partly of the
waterway formed by the Tennessee and the Cumberland, and partly of the land routes through the gaps in
the Cumberland range.   In the same manner the Kentucky settlements were born of the Ohio and the Wilderness Road.   The Cumberland and Kentucky settlements
were separated from the frontier of civilization in Virginia by a wide zone of wilderness, the rugged upland
of the Cumberland Plateau, a region condemned even
to-day by its geographical conditions to isolation, poverty, and a retarded civilization.[10]   With the exception
of Louisville at the Falls of the Ohio, the Kentucky
settlements avoided the main streams, but chose the
smaller affluents and rich undulating country between.
Filson's map (1784) shows fifty-two villages and eighteen
scattered houses.   One large group, with Lexington as
a center, spread over the land between Elkhorn Creek
and the Kentucky River and showed a few outlying
stations on the South Fork of the Licking; another
group, including Danville and Harrod's Town, was
located within easy reach of Dick's River, a southern

tributary of the Kentucky; and a third to the west stretched along the upper forks of Salt River, where Bardstown was the backwoods metropolis, and merged into the closer settlements along Beargrass Creek and the Falls of the Ohio. The geographical distribution of the early Kentucky population brings into relief the home-making motive in striking contrast to the purely trading instinct which determined the location of the French settlements across the Ohio to the north.

With the outbreak of the Revolutionary War these trans-Allegheny settlements took on a peculiar political importance. They were a wedge driven into the great West, which was claimed by the colonies on the ground of their sea-to-sea charters, claimed also by the British by the treaty of 1763 and the royal proclamation reserving the transmontane country as crown lands, and the Quebec Bill (1774) bringing the region north of the Ohio under the jurisdiction of Canada. But the Americans were on the ground. Virginia, pushed on by the pressure of land-hunger and facing the open doors along her mountain frontier, on which the best western highways converged, insisted more than any other colony upon the terms of her sea-to-sea charter. It was she who invested Clark with a commission to conquer the Illinois country, but it was this man from the backwoods settlements who proposed the scheme. The capture of Kaskaskia swept the British from the West. They had now no fort or garrison between the Lakes and the Gulf. In the face of enormous difficulty, Clark maintained his hold upon the Illinois country. The sparse population of the West afforded a scanty supply of troops to guard a frontier reaching from the Mississippi to the upper Ohio. The remoteness of the region and the barrier of the mountains interfered with the transportation of reinforcements and supplies from the East. Munitions and food came even to Fort Pitt from the Spanish at New Orleans up the Mississippi. The possession of the Illinois and Kentucky country annulled the proclamation of 1763 and the Quebec Bill, and influenced the final negotiations for peace. Though the extension of the young Republic to the Mississippi was based formally in these negotiations upon the charter bounds, the presence in these western lands of a vigorous people who had made good their title by axe and plough and rifle constituted a more solid claim to the debated territory than the yellow parchments of dead monarchs or living potentates.

NOTES TO CHAPTER IV

1. Ratzel, Politische Geographie, p. 685, and Anthropogeographie, part i. pp. 420, 424. Also E. C. Semple, Mountain Passes : A Study in Anthropogeography, Bulletin of the Amer. Geog. Soc. nos. 2 and 3. 1901.

2. Bailey Willis, The Northern Appalachians. National Geographic Monographs, vol. i. no. 6.

3. C. Willard Hayes, The Southern Appalachians. National Geographic Monographs, vol. i. no. 10.

4. Justin Winsor, The Mississippi Basin, pp. 243, 249.

5. Ibid. p. 258.

6. Evans's map made in 1755, reproduced by T. Pownall, in his Topographical Description of North America, 1776. Also Journal of Dr. Thomas Walker, in Filson Club Publications, no. 13, pp. 38–41.

7. Roosevelt, The Winning of the West, vol. i. pp. 170–172. Monette, History of the Valley of the Mississippi, vol. i. p. 313. 1846.

8. Monette, History of the Valley of the Mississippi, vol. i. p. 348.

9. Justin Winsor, The Westward Movement, p. 508.

10. E. C. Semple, The Anglo-Saxons of the Kentucky Mountains. Geographical Journal, June, 1901.

CHAPTER V

GEOGRAPHICAL ENVIRONMENT OF THE EARLY TRANS-ALLEGHENY SETTLEMENTS

THE trans-Allegheny settlements had been planted in the center of a country which comprised five eighths of the territory of the young United States. Barred from the East by the Appalachian Mountains, it sloped gently from the plateaus westward to the Mississippi and southward to the Gulf of Mexico, its surface broken only by low hills or undulating upland. A canoe could travel from Lake Erie to Mobile Bay through the heart of the region, with only two portages, one at the sources of the Wabash, and the other between the Tombigbee and the Tennessee; while on its western margin the Mississippi furnished a waterway from its northern to its southern frontier. The Ohio and its tributaries aggregate 12,000 miles of streams which in early times afforded transportation routes and facilitated the movement of men and produce.

The abundance of land, the lack of barriers, and the easy river connection, all made for expansion of western population. The vacant space between the Cumberland and the Kentucky was soon filled up; settlements were strung along the Ohio, like beads on a cord, from the Big Sandy to western Pennsylvania, while in the Northwest Territory scattered cabins up the course of the Muskingum, Scioto, Great and Little Miami rivers indicated the lines population was following.) In the frontier of a country is to be found always the index of its growth or decay. A rapid advance of the boundary, whether of settlement or political control, speaks of vigorous, abundant forces behind demanding an enlarged field of activity : a retrogression or caving-in of the frontier points to declining powers, inadequate strength.[1] Along the western waters all was activity, eager advance, yearning for a further beyond. Here appeared that uncramped, undiscouraged development which has given the distinctive stamp to American life.

The abundance of land was reflected in the generous soil of the transmontane country. North of the Ohio, the surface is underlaid by a thick mantle of glacial drift formed from rocks of a chemical nature to supply food for plants. The timberless character of the country enabled settlement to go on at a rapid rate, unretarded by the slow work of cutting down forests and clearing out stumps. South of the Ohio and beyond the rim of the Cumberland Plateau with its poor land, the blue limestone outcrops, furnishing to the subsoil an inexhaustible supply of plant food. This has made the fame of the Bluegrass region of Kentucky. It underlies also southwestern Ohio and Tennessee to the northern border of Alabama ; but owing to the covering of glacial drift in the north and the sandier character of this rock in the south, the soil is less fertile than in Kentucky.[2] The southwestern parts of Kentucky and Tennessee are underlaid by Subcarboniferous limestone, which enriches the soil almost as much as does the Silurian. The Gulf slope of this western country is a fertile silt-made plain, built up from the débris of limy or clayey rocks, and hence far richer than the crystalline detritus which covers the surface of the Atlantic plain.

An abundant rainfall well distributed through the year, a temperate climate varied by cold, dry winds which sweep down the great trough of the continent from the northwest, and warm, moist winds from the Gulf, a long, warm summer for the growing crops, all united to mitigate the hardships of pioneer life and to keep up the courage of the home-maker in the wilderness.)

The men who grew up in this westward facing country were the first genuine Americans. The seaboard population were Europeans living on American soil under English control, bred to English luxuries which were supplied by English manufactories. In time of danger English armies had been in part their resource, and the English treasury paid in part the costs of colonial wars. American soil and the barrier of the Atlantic had modified European institutions and character in the hands and persons of the colonists somewhat ;[3] but their gaze was seaward, towards the English palace and council hall where their destiny was decided. Volney, a Frenchman traveling in this country in 1796, discerned the difference of standpoint between the people of the tidewater and the over-mountain region. " The inhabitants of the Atlantic coast call the whole of this the Back Country, thus denoting their moral aspect, constantly turned towards Europe, the cradle and the focus of their interests. It was a singular, though natural

circumstance, that I had scarcely crossed the Alleghanies, before I heard the borderers of the great
Kanhaway and the Ohio give in their turn the name of
Back Country to the Atlantic coast, which shows that
their geographical situation has given their views and
interests a new direction, conformable to that of the
waters which afford them means of conveyance towards
the Gulf of Mexico."[4]

In the cabin clearings of the western wilderness, beyond the barrier of the mountains, English institutions
took on a new stamp of republicanism, society became
more purely democratic, and the new-born American
looked only to his own strong arm for aid, to his own
strong intellect for counsel. Separation from old-world
traditions, a return to close contact with nature, the
stripping off of non-essentials, growth under conditions
of uncramped development and untried possibilities —
these made the sturdy, youthful American of the western
wilds. Old-world methods and ideas must be transformed to survive in new-world conditions,[5] and the conditions modify the man along with his methods. The
change may appear at first to be retrogression, but it is
only the step backward for the long running-jump.

The successful ending of the Revolutionary War had
focused the eyes of the Americans on their own country.
Then the possibilities of the young West began to loom
up, especially in view of the economic depression of the
tidewater region following the long war. Warrants and
military scrip for western lands guided thither the dis-
charged soldier from New England to South Carolina.
The outbound population to the Kentucky and Cumber-

land became more composite in its character than that of
the old backwoods on the Piedmont and Blue Ridge
frontier. It included the same racial elements and in
addition more varied American constituents, which beyond the mountains forgot their colonial allegiance,
whether to the aristocratic capital of Charleston or
the Puritan township of Connecticut, lost their old sectional feelings, and instead substituted allegiance to the
national government and the sectionalism of the over-
mountain men as opposed to the seaboard states.

Upon these early western settlements isolation set its
stamp. Range after range of mountains, mile after mile
of rugged plateau separated them from the seats of civilization and government. Land and water routes were
beset by Indians, and even the homeward-leading rivers
beat back with their increasing currents the pirogue of
the eastbound Westerner. The great streams of the Mississippi valley made the settlements more accessible to
their British neighbors to the north, and their Spanish
neighbors to the west and south ; but this accessibility
was disadvantageous rather than otherwise. Detroit
and the navigable waterway of the St. Lawrence and the
Great Lakes played the same part in the history of the
young West as did New Orleans and the Mississippi.
Both cities were the center of plots to incite the Indians
— northwest and southwest — against the backwoods
frontier, to construct the intervening savage tribes into
buffer states against the expansion and aggression of the
colonists, to invade American territory for illicit trade,
and finally to tempt the frontiersmen to defection by
promises of the needed river outlet for their produce to

the sea. By the retention of the Lake posts for twelve
years after the peace of 1783, the British kept their hold
on the fur trade in the Northwest Territory. Moreover,
they secured permission from the Spanish to trade on the
western bank of the upper Mississippi, and this necessitated their passing to and fro on American soil by way
of the Chicago portage or by the Wisconsin River to the
post at Prairie du Chien. In consequence of this tres-
passing the American trade at Vincennes was greatly
reduced.[6]

From the south, the Spanish traded up the Alabama
and Tombigbee rivers from Mobile across the thirty-first
degree, north latitude, which had been agreed upon in
the treaty of 1783 as the northern boundary of West
Florida, and over the three-mile portage to the basin of
the Tennessee ; and traders from Pensacola were doing
a thriving business with the Creeks and Cherokees to
the northeast. From New Orleans and other Spanish
posts along the Mississippi, traders were encroaching
on American territory, especially up the Illinois River
where they shared the field with Canadian rivals, while
the officials on the Gulf were obstructing or inhibiting
the navigation of the Mississippi to the western pioneers
and hounding on the southern savages to attack.

On either side of the western wedge of settlement lay
a broad zone of Indians, whom the expansion and encroachments of the whites, as well as the incitement supplied by British and Spanish, made the natural foe of the
settler. They were as much a part of his environment
as the fertile soil and the far-reaching wilderness. They
controlled in part his social organization, taught him new

modes of warfare, and modified his character. A frontier is never a line but always a shifting zone of assimilation, where an amalgamation of races, manners, institutions, and morals, more or less complete, takes place.[7]
The English pioneers in the wilderness retained their
sedentary occupation of the land, in contrast to the
nomadic habits of the French traders; settled in more
or less strongly compacted groups, which were further
consolidated locally and politically by the danger of
the all-surrounding savage; and thus retained an inner
environment which was civilized. The line between
them and the savage was therefore strictly drawn : half-
breeds were rare. But the outer environment was all of
the wilderness and the Indian, and to this the man of the
Cumberland and Kentucky yielded himself. He lived
in great part by the chase, dressed in buckskin and furs,
wore the moccasins of the redman, adopted his scalping-
knife and tomahawk, and waged against him a war of
extermination, with all its savage features of ambush
and scalping, and all its brutalizing effects.[8] In view
of the ever-present danger of Indian attack, the basal
organization of pioneer communities was military. The
settlement was the stockade or station, — a fortified village; militia service in the common defense became the
first duty of the citizen. Representation in the early
Kentucky conventions to consider separation from Virginia was made on the basis of military companies.

The common remoteness and the conditions of wilderness life laid their equalizing touch upon all. Equality of opportunity and resource, identity of tasks and of
dangers, and the simplicity imposed upon all precluded

classes, and in the mass developed vigor, enterprise, and independence. The backwoodsman had what the forests and clearings could furnish, and little more. Wooden vessels of all kinds, whether turned or coopered, were in common use. Owing to the great cost of transportation over the mountains, hardware was rare. Houses were built without nails, and even shingles were put on with oak pins.[9] Linen was made from the lint of nettles, and buffalo wool was the raw material for cloth. Furs became the medium of exchange, though after the Revolution " there was some paper money in the country which had not depreciated more than one half, as it had at the seat of Government."[10] This fact speaks eloquently for the isolation of the region. At first salt was imported at almost prohibitive prices; but soon the pioneer began to boil water from the saline springs which abounded, and later learned to bore for richer brine. At Bullitt's Lick on Salt River in Kentucky, a regular industry was started, and supplied the frontier communities from the Ohio to the Cumberland.

The products of the country which were bartered for eastern merchandise were primarily those of the wilderness — hides, furs, ginseng, snakeroot and bear's-grease; or of a frontier country with abundant pasture land or forest range, such as horses, hogs, salted pork, lard, tallow, and dried beef. Very soon the fertile soil began to yield abundant crops. Tobacco, corn, flour, whiskey, flax, and rope from the hemp-fields, were ready for export. Before the treaty of 1783, when Spain was friendly, much of this produce went down the Mississippi and found a ready market in New Orleans.

Only the most valuable part would bear the cost of up-stream and over-mountain transportation to the eastern seaboard.

The high price of commodities imported by this route forced the backwoods settlements to institute manufactures at a surprisingly early date, so that here in one small area were to be seen all the stages of economic development, — savage, pastoral, agricultural, and industrial. The line of manufactures was determined by the domestic supply of raw materials and the most pressing need of the settlers. Retarded by the selfish policy of England, manufactures along the seaboard scarcely antedated the appearance of industries in this free western country. Iron was found in great abundance on the westward slope of the Allegheny Plateau along the Monongahela, Youghiogeny, and Cheat rivers, and the big demand for it along the frontier caused the erection of furnaces and iron works as early as 1788.[11] Lexington, Kentucky, had a cut-nail factory in 1801; and Georgetown a ropewalk and fulling-mill in 1789.[9]

The first spinning-jenny after the Hargreave type in America was operated at Philadelphia in 1775; a cotton factory was established at Beverly, Massachusetts, in 1787, at Providence in 1788, and in Pawtucket, Rhode Island, in 1790.[12] A society for the encouragement of manufactures was organized in Danville, Kentucky, in 1789, under the auspices of Judge Harry Innes and other prominent men, who in 1790 started the first cotton factory in the West. The carding, spinning, and weaving machines were purchased in

Philadelphia, where they were manufactured, carried over the mountains at great expense to Pittsburg, and down the Ohio to Louisville, whence they were hauled to Danville. The raw cotton was produced in the Cumberland River settlements, and for several years was a regular article of trade between the two communities; some was also raised in Kentucky. The four skilled workmen and the manager were imported from Philadelphia, and John Brown, congressman from the Kentucky district of Virginia, who negotiated the purchase of the machines and the contract with the workmen, explained to Judge Innes, in a letter of May 25, 1790, that a certain irritating delay would never have occurred "did not a disposition prevail with some in this place to prevent the emigration of Manufacturers and the establishment of Factories in the Western country." [13] Here speaks the new sectionalism. Already some of the wiser heads in the East had predicted that, if the federal government failed to secure by treaty with Spain the free navigation of the Mississippi for the transmontane settlers and an outlet for their agricultural products, these communities would be early forced into industrial ventures, and so rendered economically independent of the northern seaboard states, who hoped to find in them a market, and yet, for selfish reasons, desired the closure of the Mississippi.

Nature had done everything to open up Kentucky, Tennessee, and the growing Northwest Territory towards the south. Ramifying through this whole country was the Ohio with its tributaries, affording easy access to the Mississippi and the natural market in semi-tropical

New Orleans. Old established routes of trade passed their door ; and the reckless, jovial French voyageur, with his gaudy backwoods dress and his fiddle, was still passing down the Wabash or Miami from the Lake country to trade at Louisville, St. Louis, and New Orleans.) With the cessation of the British and Indian wars, and with the increasing tide of immigration, the Ohio route from the East grew in importance. Brownville, laid out at Redstone Old Fort on the Monongahela in 1785, and Pittsburg as the starting-points for the downstream journey, became lively distributing-centers for the western country. In both places an active boat-building industry supplied boats for the hundreds of emigrants who arrived every week from the East, and also for the regular trade on the river. Kentucky offered more to settlers than the country across the Ohio ; and now they began to fill the northern part of the state which had been left vacant because of its exposure to the attacks of the Shawnees from the Scioto. Now the old war-trail of the savages along the outer rim of the Cumberland Plateau from the mouth of Limestone Creek to the Kentucky River settlements became a highway of trade. Limestone became the point of debarkation of men and goods for central Kentucky, and four miles back from the Ohio, on the top of the steep ridge facing the river, grew up the town of Washington. Here large wagons for the interior started with loads too heavy to be hauled up the steep slope. The Falls of the Ohio made Louisville a natural port for the upper river, as also the head of navigation for the lower stream. The town was furthermore the western terminus of the Wilderness Road. Hence it afforded unusual facilities for trade.[14]

In those early days commerce moved in wide circles. The western pioneer from Nashville or Brownville passed down the Cumberland or the Monongahela with his own produce in his own boats to the Ohio, thence down the Mississippi to New Orleans, where his cargo was sold or exchanged. From there he went by sea to Cuba for further sales, and then embarked for New York, Philadelphia, or Baltimore, where his money was reinvested in manufactured articles. These he transported over the mountains at a cost of three dollars per hundredweight, to sell them in the scattered markets of the Ohio basin, after having been absent from four to six months. The result of such circular tours was large financial profit and a broad experience of life which made these men of the backwoods in a sense the cosmopolitans of the country. Smaller boats, laden with fine commodities of slight bulk and great value, sometimes made the return trip up the Mississippi from New Orleans to Louisville. This took forty days or more. The barges, which were equipped with sails and oars, were helped on by the prevailing south winds and the eddying up-currents at the bends of the river.

For the first decade of their history, the Kentucky and Cumberland settlements relied chiefly on Pittsburg and the Ohio for the limited trade they could furnish or command ; but as the population increased, and the rich fields began to respond to systematic culture, New Orleans was looked to as the natural market for western products, and as a source of supply for foreign and domestic products not to be secured nearer home. Therefore, when at the close of the Revolution Spain laid claim to both sides of the Mississippi River as far north as Kentucky, to the exclusive control of its navigation, and to the right to impose transit and harbor duties; and when in 1787 Jay, the secretary of foreign affairs, was about to make a treaty with Spain conceding this right, bitter protests came from the transmontane settlements. In view of the increasing trade the barrier of the Appalachians made the outlet of the Mississippi a vital question. Collot in 1796, computing the expense of transportation by the Potomac and other over-mountain routes as compared with the Mississippi, found that goods could be conveyed from Philadelphia to Kentucky at a cost of 33 per cent. *ad valorem,* and from New Orleans to the Illinois of only 4 to $4\frac{1}{2}$ per cent.[15] Moreover, the difference in climate and consequently of products between Louisiana and the Ohio country made each a market for the other; and the Gulf region, with that lack of diversification of products which characterizes all new countries, was confining itself to cotton, sugar, and molasses. From 1785 to 1795 the exorbitant duties on American commerce descending the Mississippi and the oppressive commercial regulations imposed by the Spanish brought stagnation on the river, while in the frontier settlements from the Tennessee to the Allegheny it was the moving cause in the separatist movement. But other causes were operating also.

Along this whole western frontier appeared that tendency towards defection which we have found to be characteristic of all peripheral holdings. The causes lay in the remoteness of the settlements, the barrier of the Appalachians, and the closure of the western outlet. Geographical conditions had produced here a people with different point of view, needs, and interests from those of the seaboard.) Their situation reproduced in essential features that of the colonists before the Revolution.   In the constantly recurring Indian wars, troops and munitions from the tidewater capitals always arrived late, delay was dangerous, and the frontiersmen had to act on their own responsibility.   The East did not appreciate the attitude of the frontier towards the Indian : the frontier saw the futility of Indian treaties, blamed the national authorities for talking and bribing instead of fighting while American settlements were being ravaged by the savages, and resented the course of the national agents in ignoring state and private agreements with the natives.   Peace came to the frontier only when a punitive expedition from the Holston or Cumberland or Kentucky or Ohio, unauthorized by state or national authorities, used the only arguments appreciated by the Indians.   The bloody raids of the savages in 1786 called forth reprisals when George Rogers Clark, Benjamin Logan, and Simon Kenton led expeditions against the Shawnees north of the Ohio, and Robertson with a hundred and thirty hardy followers sallied out from the Cumberland settlement to visit retribution on some marauding Creeks and Cherokees who had intrenched themselves near the bend of the Tennessee.   Requisition was made upon the settlers for pack-horses and supplies, ranging from salt to lead, with which to equip these forces.

Such expeditions, being of the nature of private enterprises, did not have their expenses defrayed by the government. The costs fell, therefore, as an unequally distributed charge upon the communities, who felt in consequence the injustice of being taxed while their effective and constant military service went unpaid. The excise taxes were a special grievance and led to the Whiskey Rebellion in western Pennsylvania, and to milder protests elsewhere along the transmontane frontier. The surplus corn in the remote settlements, being of large bulk in relation to its value, could not stand the heavy cost of transportation across the mountains until it had been converted into whiskey. Here a nature-made law was confronting a human one, as it continues to do in the isolated regions of the southern Appalachians to-day. In Pennsylvania the law was resisted as unjust, and in the mountains to-day it is evaded by the moonshine still. Similar geographical conditions produce similar results. In other respects the burden of distant tidewater control rested heavily upon the frontiersmen. They had their local courts for minor cases, but all the more important civil and criminal cases and all appealed suits had to be carried to the tidewater capitals for trial. This involved a journey of from three to five hundred miles, much expense, and long delay. On account of courts and taxes, therefore, the pioneers desired separate state governments. But most of all they desired the representation in Congress which would give them a voice in the pending negotiations with Spain regarding the navigation of the Mississippi.

All along the frontier from western Pennsylvania to

the Tennessee River the spirit of independent statehood was rife.[16] Some of the frontier communities wished to establish their states with boundaries geographically determined, ignoring the charter lines of the old colonies. They were guided by the principle of ready communication. The Watauga settlements in western North Carolina proposed to combine with the Virginia settlements further up the valley of the Holston and along the New, because together they formed a geographical whole. As Nashville, through the river and the southern branch of the Wilderness Road from Crab Orchard, stood in closer communication with her northern than her eastern neighbors, there was talk of the union of the Cumberland with the Kentucky settlements.

Blocked by the Eastern states in their efforts to secure independent statehood, disgusted by the short-sighted policy of the federal government which threatened to sacrifice the interest of the over-mountain communities in the free navigation of the Mississippi, and exasperated by its desultory efforts to stand on its rights with Spain, while the commerce of the Mississippi was being ruined by the depredations of Spanish officials on vessels which attempted to descend the river without their permission, these pioneer settlements finally developed a separatist movement of a different nature, which aimed at union with any foreign power, Spanish or British, promising to give them an outlet to the sea. The temptation was very near, and it was geographically determined.

Every river system forms an unbroken whole and therefore serves as a natural bond of union between those living among its remotest sources and those settled

at its mouth.[17] The direction of its flow guides the drift
of commercial intercourse and the trend of political combination.   Down the course of the Dnieper and the
Volga moved Russian trade and Muscovite dominion.
The Danube carried Roman sway to the Black Sea.
The most serious wars of Holland had their cause in
the desire of France to control the mouth of the Rhine.
For twenty years the politics of our western country
centered about "the Island of New Orleans."   The
pioneers were under the geographical control of the
western waters.   Their rivers carried them on downward currents to the Mississippi and the Gulf, where
the Spanish offered free navigation and trade.   From
the Ohio the winding courses of its northern tributaries
carried them to easy portages leading to the vast waterway of the Lakes and the St. Lawrence.   Between them
and their blundering Congress in the East stretched a
mountain wall three hundred miles wide.   In the colder,
undeveloped north they could find a near market, as in
the warmer, half-developed south, while the market to
the east was cut off by a mountain barrier.   There was
another alternative, rather ideal in its character but
favored by geographical conditions : union with the
British and a downstream conquest of New Orleans.
This possibility was held over the head of the Spanish
Intendant on the Gulf.   Spanish influence was felt more
in Tennessee and Kentucky, British in Kentucky and
the infant settlements of the Northwest Territory, where
the Muskingum country had its Colonel Wilkinson in
General Parsons.

All such intrigues ceased with the promulgation of

Pinckney's treaty in 1795 which secured the free and unlimited navigation of the Mississippi, with the right of deposit at New Orleans free from Spanish control. The ultimate result of the Mississippi outlet, the discussion it aroused, and the benefits it bestowed, was the purchase of Louisiana eight years later.

NOTES TO CHAPTER V

1. Ratzel, Anthropogeographie, part i. pp. 260–263.

2. Shaler, United States of America, vol. i. pp. 105, 106.

3. F. J. Turner, The Significance of the Frontier in American History, Annual Report of the Amer. Hist. Ass'n for 1893.

4. C. F. Volney, View of the Climate and Soil of the United States of America, p. 21.   London, 1804.

5. Roscher, Nationalökonomik des Ackerbaues, p. 104.   1888.

6. Justin Winsor, The Westward Movement, p. 416.

7. Ratzel, Anthropogeographie, part i. p. 265.   1899.

8. Monette, History of the Valley of the Mississippi, vol. ii. chap. 1. 1846.

9. R. H. Collins, History of Kentucky, vol. i. p. 516.

10. Dr. Humphrey Marshall, History of Kentucky, vol. i. p. 150.   1812.

11. Monette, History of the Valley of the Mississippi, vol. ii. p. 199.

12. Weeden, Economic and Social History of New England, vol. ii. p. 848.

13. George W. Todd, The First Cotton Factory in the West, an unpublished paper read before the Filson Club, March, 1898, based upon unpublished letters and papers of Judge Harry Innes, great-grandfather of Mr. Todd.

14. Louisville : A Study in Economic Geography, Journal of School Geography, Dec. 1900.

15. Justin Winsor, The Westward Movement, p. 508.

16. F. J. Turner, Western State-Making in the Revolutionary Era, Amer. Hist. Review, vol. i. nos. 1 and 2.

17. Ratzel, Anthropogeographie, part i. pp. 344, 345.

CHAPTER VI

THE LOUISIANA PURCHASE IN THE LIGHT OF GEOGRAPHIC
CONDITIONS

FOR twenty years after the treaty of 1783 the Mississippi was accepted as the western boundary of the
United States. But a river is not a barrier and therefore never a scientific boundary ; it is merely a convenient line of demarcation,[1] uncertain at best on account
of its shifting bed. The plantations on the meanders
of the lower Mississippi belong now to one, now to the
other of the contiguous states, as the mighty stream
straightens its course after the almost annual overflow.
The Rio Grande is anything but a satisfactory boundary
between the United States and Mexico. Dry for many
months of the year, it bears no semblance of a barrier ;
then is suddenly filled with a resistless current which
cuts a new channel for itself, leaving many square miles
of Mexican territory on its northern bank, and in turn
cutting off Texas ranchers from their native state.
The result is a new survey or endless controversies as
to whether Texas or Mexico is to claim certain dislodged
pieces of land.

The political frontier line which is run along a river
is an artificial one, for every drainage system forms an
unbroken whole. The common destination of the tributary streams makes it as easy to cross as to follow the

structural axis of the basin. The water journey on the Mississippi from Cairo, Illinois, to St. Paul, Minnesota, and return presents about the same difficulty as the round trip from Cincinnati to Kansas City by way of the Ohio and Missouri. To migration and to conquest a river presents no obstacle. Within the Tigris-Euphrates valley three great monarchies — Assyrian, Babylonian, and Persian — arose, and each in turn extended its dominion over the whole basin. To the Romans, the Danube and Rhine as their northern frontier had the value chiefly of established lines in the unexplored wilderness of the interior ; in a minor degree, also, of strategic positions in defensive warfare for a well trained army, like that of the Tugela River in the recent Transvaal War. But even the Romans found it necessary to acquire the highlands forming the watershed between the Rhine and the Neckar, and Dacia to the north of the Danube. The Danube formed the northern Turkish frontier in 1444, but in twenty years the empire had taken in Wallachia across the river and by 1672 embraced the whole basin of the Danube up beyond Buda-Pest. When the empire began to decline, the frontier dropped back down the river to Belgrade (1700), but still held both sides of the valley. It is safe to predict that the Oxus will not long remain the boundary of Bokhara in the hands of Russia, but that its southern tributaries will become so many wedges of Russian expansion to split off blocks of Afghanistan territory.

Even more surely than conquest does migration tend to hold the two slopes of a river basin. Before the westward movement of the Germans, both banks of the Rhine were held by the Gauls, whose villages in some instances were bisected by the river, just as Rome arose on both sides of the Tiber. Then in a like manner the Teutons spread over the whole long, deep trough of the Rhine valley. The trans-Rhenish provinces were lost to Germany in the eighteenth century by the aggressive policy of France, but 1871 saw the restoration of the natural boundary along the crest of the Vosges Mountains. Across the Rhine, Meuse, and Seine the early Franks spread into northern Gaul, and they occupied the valley of the Rhone through the pass of Belfort. The population of all these river basins show the admixture of that tall, blond race, while the surrounding highlands harbor the darker, shorter race constituting the earlier inhabitants of the land.[2]

A boundary of race or language cuts across the axis of a river basin; rarely is it defined by the stream itself, and then only for a short distance. Where the Elbe flows through the low plains of North Germany, across its whole valley the population is the pure Teutonic type — fair, tall, long-headed; a broad zone of a more brunette type occupies its middle course across the uplands of Saxony; while its upper course, hemmed in by the Erz and Riesen Mountains, shows the broad-headed, short people of the Bohemian plateau.[3] Lines of ethnical demarcation, therefore, cut the Elbe valley transversely, not longitudinally. The whole valley of the Danube from the Drave northward to the Austrian boundary is Hungarian. From the Drave mouth to the Iron Gate it is Serbo-Crotian on either bank, as is also its western tributary, the Save. The linguistic

boundary of German cuts across the Danube between
Vienna and Buda-Pest.[4]   The German-Romansch line
of speech crosses the upper Inn at the entrance to the
Engadine, and the upper Rhine near Ilanz, some twenty
miles above Chur.[5]   So in America, the lower St. Lawrence is French and the upper stream is English.   This
linguistic boundary is one of the few scars on a political body to be found in this country ; such scars, being
made in general in the youth of our nation, soon healed
over and were obliterated.   The boundary line of
the Mississippi was never more than a scratch at best,
as we shall see, for as a race frontier it was only an im-
aginary line.   As soon as the enterprising Westerners
reached the Mississippi, they were on it and across it.

The trans-Allegheny rivers made the requisite conditions for a rapid expansion of population, and the
convergence of the Ohio, Tennessee, and Cumberland
upon the Mississippi guided the three incoming tides of
settlers towards the great central stream.   As early as
1788–89, between eight and nine hundred boats went
down the Ohio past Fort Harmar (Marietta), carrying
20,000 people, with 7000 horses, 3000 cows, 900 sheep,
and 600 wagons.[6]   This does not take into account the
trains of pioneers whose feet were treading out a broader
trail along the Wilderness Road, or the large accessions to the Cumberland settlements after the cessation
of the Indian wars.   The numbers of the western people were ominous to the Spanish power beyond the Mississippi ; more ominous still were their spacial ideas.
Already the bigness of the West was being absorbed
into their mental constitution.   The mighty sweep of

the western waters, the interminable stretch of the northwestern prairies, the boundless extent of the Appalachian forests, were giving to these men from the narrow Atlantic plain new standards of measure. Five to ten thousand acres became the desirable, almost the usual size of a farm, until the land was pretty well taken up. A river voyage of a thousand miles was a commonplace matter, and even a land journey of five hundred miles on foot or horseback along the Chickasaw Trace from Fort Adams, the southernmost post of the United States on the Mississippi, to Nashville was nothing to brag of. It was made every year by the returning traders with their money from the river trips to New Orleans.

In 1800 the western settlements extended down the Ohio almost as far as the mouth of the Cumberland. There were besides two remote outlying strips of settlement along the Mississippi, the form and location of which were significant. The northern one was in the western part of the present state of Illinois and formed a narrow band along the Mississippi from near the mouth of the Illinois River southward beyond the Kaskaskia. Its nuclei were to be found in the old French settlements of Cahokia and Kaskaskia, and the attraction was the opportunity of trade with St. Louis and Ste. Genevieve on the Spanish side of the river, where could be obtained foreign articles of luxury from New Orleans or skins and furs which found their way down from the upper Missouri.

The other group was known as the Natchez District and stretched along the bluffs of the Mississippi for a hundred miles between the Spanish line of demarcation and the Yazoo delta. Its population of about 6000, including slaves, was distributed in several large settlements along the Bayou Pierre, Cole's Creek, St. Catherine, Second Creek, Honochitto, Buffalo Creek, and Big Black River, most of them within ten or fifteen miles of the Mississippi. The District contained a nucleus of English settlement, dating from the period of 1763–1783, when the British held West Florida and were trying to divert some of the commerce of the Mississippi by way of the Iberville River, which was to be canalled, and Lake Ponchartrain to Mobile or Pensacola.[7] There were new accessions of American population after the treaty of 1783, which secured the region to the United States, though the Spanish did not yield the control which they had assumed by encroachment till 1798. The proximity of Natchez to the Spanish boundary made it the most convenient point for trade with New Orleans, and it was the starting-point of the old wilderness trail to the Cumberland settlements for the returning river men. The bluffs above the Mississippi afforded a more hygienic and less malarial situation than the low bottom-lands either above or below. The elements of the population were characteristically those of a frontier region in this locality — Anglo-Americans from the old British Florida, a few French who had become dissatisfied with Spanish rule in Louisiana, a remnant of Spaniards, and an increasing proportion of Westerners.

The nature of a frontier as a zone of commingling elements has been described above. The composite character of the Natchez population was duplicated on the west side of the Mississippi with the proportions of the Latin and Saxon constituents reversed. In the barge and flatboat life on the mighty stream itself, the two races were in all probability about equally represented. Not more completely did the Ohio and the Missouri mingle their currents in the central channel than their burdens of craft and men and merchandise. Here was the cosmopolitan life of the country — river men from the Holston and the Allegheny, newly arrived New Englanders from the Muskingum settlements, French, British, and Scotch traders from the Illinois River, and French voyageurs with a few Spaniards, chiefly officials, from the Louisiana side. Busy was the trade on the river. By it New Orleans and Brownville were linked together. The Creole city drew its flour, grain, and salted meat from the Ohio country, its vegetables from the Natchez District, so that when suspension of the American right of deposit at New Orleans put a quietus on the river traffic, it nearly starved.

In Upper Louisiana, French and American elements predominated in the long strip of settlements which bordered the Mississippi from St. Charles and St. Louis to New Madrid. Some of the French had belonged originally to the east side of the river, and had migrated from the Wabash and Illinois when this region was acquired by England in 1763. American settlers were tempted across to New Madrid by Governor Miro of Louisiana in 1787 by large grants of land and promises of free navigation of the Mississippi, and under the same allurements many from the Holston and Cumberland emigrated to West Florida. By the census of

1799, the population of St. Louis was 925, St. Charles 875, Ste. Genevieve 949, New Bourbon 560, Cape Girardeau 521, and New Madrid 782, all of them river towns. The fact that Upper Louisiana had increased its population much more rapidly than the remainder of the province[8] was due largely to the steady influx of American colonists. A regular extra-territorial expansion was going on : the backwoodsmen were anticipating the movement of the flag : the race, by an anthropogeographical law, was spreading from the downward to the upward slope of the river valley.

The trans-Mississippi country had a stronger attraction for the pioneer than the region north of the Ohio ; the climate and soil were better adapted to the crops which he was accustomed to raise with slave labor, and there slavery was not forbidden. Somewhere between 1797 and 1799, Daniel Boone, having lost his lands in Kentucky by a defective title, moved to Spanish territory and took up his residence in the Femme Osage settlement, about forty-five miles west of St. Louis. His son, Daniel Boone, Jr., had preceded him thither, and other of his married children soon followed him.[9] Captain Meriwether Lewis, in the spring of 1804, found in this settlement thirty or forty families from the United States, and another smaller group of American emigrants on the Bon Homme Creek a few miles to the east.[10] All the river towns were getting accessions from the eastern side of the Mississippi.

Lower Louisiana was more remote from the American frontier of continuous settlement, and hence its population showed a smaller admixture of the American element. This was found chiefly in New Orleans, where the advantage of trade attracted the enterprising merchant from the United States. There was an occasional American planter between the Mississippi and the Washita, and some American immigrants at Natchitoches far up the Red River; while a band of adventurers, under Philip Nolan, in 1800 penetrated to the Brazos River in the present state of Texas. There they had set up their cabins, and were occupied in catching wild horses, till they themselves were captured in 1801 by Spanish officials.[11]

The zone of assimilation which constituted the American-Spanish frontier was shifting ominously to the western side of the political line of demarcation. Louisiana was in danger of being inundated by the on-coming flood of western settlers. The Intendant at New Orleans became alarmed, and in 1799 issued regulations to discourage American immigration in every way; and in 1802 the king of Spain prohibited any grant of land to a citizen of the United States.[12]

The expansion of the Westerners and the oneness of the great river valley were working together for the acquisition of Louisiana by the United States. Race power and geographical condition were operating to the same end. Jefferson understood this. In January, 1803, he urged upon Congress the desirability of purchasing from the Chickasaw Indians a strip of territory along the Mississippi, north of the Yazoo, in order to possess " a respectable breadth of country on that river [the Mississippi] from our southern limit to the Illinois at least, so that we present as firm a front on that as on our eastern border." [13]   And again, he affirms his faith in the policy of putting off the day of contending for the possession of New Orleans " till we are stronger in ourselves and stronger in allies, but especially till we have planted such a population on the Mississippi as will be able to do their own business." [14]

Historians are fond of insisting that the acquisition of the Louisiana country was fortuitous, depending upon half a dozen chances.   The great factors of a nation's development are not set in operation by the whim of a First Consul or the uncertain events of a distant war. The purchase of Louisiana was the occasion, not the cause of the acquisition of the trans-Mississippi country. That must have come sooner or later.   Even if the French had established themselves in Louisiana, they could not long have resisted the operation of geographic factors and the enterprising spirit of the western people, itself in part a product of environment.   The trans-Mississippi region, hopelessly arid beyond the one-hundredth meridian, could never have supported a large enough population to resist the Americans, with whom the common navigation of the Mississippi would soon have brought them to blows.   At the moment negotiations for the purchase were pending, the breadth of the Atlantic rendered formidable an uprising of the blacks in San Domingo and lost France that island.   Napoleon applied the geographical principle involved to a possible French colony in Louisiana, further jeopardized by a foreign war.   Had England conquered Louisiana from France — the chance which Napoleon feared — even her superior colonizing methods could not have made the country support a population large enough to cope with the thickly planted American settlements in the wide, rich, well watered regions to the east. In a conflict between a cis-Mississippi and a trans-Mississippi power, the former had every geographical condition in its favor, — coast-line, rivers, climate, soil, and habitable area. The Americans were destined to hold the West. The purchase hastened and facilitated the process.

For the time being, the United States could not have had safer neighbors in the Louisiana country than the Spanish. They were too weak to be aggressive, and too unprogressive to offer serious competition to American trade. Jefferson's only fear was lest they should be too feeble to hold the country " until our population can be sufficiently advanced to gain it from them piece by piece." The arbitrary suspension of the right of deposit at New Orleans in 1788, and again in 1802 sufficed to indicate the annoyance and hardship to which the Americans would be exposed so long as the key to the western commerce was in any hands but their own. The excitement over this violation of the treaty rights was greatest in the West, and here it was tremendous. Kentucky and Tennessee in 1799 could with difficulty be restrained by the federal authorities from sending out an independent expedition to capture New Orleans. Here spoke the self-reliant, dauntless spirit of the backwoods. In 1802, the President was flooded with protests and memorials from the whole transmontane country. Stirred even more by the rumors of the cession of Louisiana to France, Kentucky and Tennessee called for immediate action, and Governor Claiborne of the

Mississippi Territory offered to raise in his district alone a sufficient number of volunteers to seize New Orleans before it should be transferred to French hands.

The prospect of having these new neighbors in the Louisiana country was not reassuring. The French were at this time the mightiest nation in Europe. Their recent training had not made them pacific, and the restless, energetic, impulsive temperament of the people was sure to produce friction when they and the Americans met on the waters of the Mississippi. "There is on the globe one single spot," said Jefferson, "the possessor of which is our natural and habitual enemy. It is New Orleans, through which the produce of three eighths of our territory must pass to market, and from its fertility it will, ere long, yield more than half of our whole produce and contain more than half our inhabitants. France placing herself in that door assumes to us the attitude of defiance. . . . The day that France takes possession of N. Orleans fixes the sentence which is to restrain her forever within her low-water mark. It seals the union of two nations who in conjunction can maintain exclusive possession of the ocean. From that moment we must marry ourselves to the British fleet and nation." [15] These measures were not desired, but would be forced upon the United States " as necessarily as any other cause, by the laws of nature, brings on its necessary effect." If the French insisted upon holding Louisiana, the cession of the Floridas and the Island of New Orleans might reconcile the Americans by guaranteeing their free access to the Gulf of Mexico. The transit duties on all commerce passing in or out of

Mobile Bay also had been working a hardship upon the small Tombigbee settlements, but the key to the Mississippi was the crucial thing.

Napoleon's dreams of colonial empire had to be relinquished. The rebellion of the negroes in San Domingo demonstrated the futility of oversea domination for France at this time. British naval power and the British army in Canada made evident the impossibility of holding Louisiana in the event of war with England. French agents represented to Napoleon the great difficulties of an upstream campaign from New Orleans to protect the distant settlements in Upper Louisiana against the aggressions of the Americans, and the ease of a downstream conquest of Lower Louisiana by the determined Westerners. To the insight of the First Consul into geographical situations, it was clearly futile to retain Louisiana and relinquish New Orleans, which was the key to the whole province. But the dominant factors in the alienation of this western empire was the broad Atlantic and the island-born naval power of the English. Thus, in every motive for the sale as for the purchase of Louisiana, the geographic factor were the strongest.

The retrocession of Louisiana to France had been made by Spain reluctantly, but the presence of a strong buffer state between Mexico and the expanding United States was undoubtedly a compensation.[16] With the annexation of Louisiana to American territory, however, the situation of Spanish holdings in the northern continent was seriously affected. Nearly the whole sweep of the Gulf and Caribbean coast from the Guianas around to Porto Rico was in the hands of Spain, except the five-hundred mile stretch of the Louisiana littoral between the mouth of the Sabine and the Mississippi; but this possessed an importance out of all proportion to its length. The significance of a coast is determined largely by the extent and character of its hinterland. Back of the Louisiana coast lay the vast Mississippi basin, now wholly American, like a great funnel whose mouth was at New Orleans. The pressure from the back country was sure to be enormous and eventually to burst the bonds of Spanish restriction on the Gulf coast. The Spanish holdings in Florida and Mexico were now severed, and that at a strategic point. If Louis XIV., from his base in Canada, had succeeded in his plan of conquering the Hudson valley and intercluding the two lines of English coast in America, the politico geographical situation would have been analogous.

According to the Purchase treaty, as understood in the United States,[17] the Louisiana province was to comprise the original French holding from the Rio Grande del Norte to the Perdido River, which defines the western boundary of the present state of Florida. But Spain, with the instinct of self-preservation, tried to restrict these limits as much as possible, transferred only that part of the Louisiana coast lying between the Sabine and the Mississippi, together with the Island of New Orleans on the east bank of the river, planted new colonies in eastern Texas and new posts along this frontier, and began to encroach with an armed force along the Sabine and the Red, with the secret purpose of making the Red River the eastern boundary of Mexico[18] and thus excluding the United States from the Gulf except at New Orleans.

As well might they have tried to dam the mouth of the Mississippi. Rivers and nations strive equally to reach the sea. Soon after the occupation of Louisiana by the United States, its Gulf coast was stretched eastward to the Pearl River. What was known as the Baton Rouge District of West Florida had a population made up largely of English, dating from the British supremacy in this region, and Americans who had come in from the Cumberland country and the Natchez settlement near by. A revolt in the Baton Rouge District and an appeal to the United States for annexation to the Territory of New Orleans brought the American flag to the Pearl River. The exigencies of war carried it to the Perdido, when England, countenanced by Spain, undertook to use the Florida coast as a military base during the war of 1812,[19] and Congress, to defeat this project, ordered its occupation and annexation to the Territory of Mississippi. Purchase secured East Florida, and expansion of American settlers to the southwest finally extended the Gulf coast of the United States to the Rio Grande.

The extension of the federal authority over the province of Louisiana was the signal for an inrush of American settlers, both from the eastern and the western states. Six months before the transfer hundreds had collected on the Cumberland for the purpose. The distribution of population in 1815 in the New Orleans Territory, which comprised the present state of Louisiana, is instructive as illustrating the geographic control of watercourses. The greater portion of the 90,000 inhabitants was found along the Mississippi for seventy

miles above and thirty miles below New Orleans. Most
of these were Creole French, with a slight sprinkling
of Americans.   There was a dense French population
on the Bayou Lefourche for fifty miles below its efflux

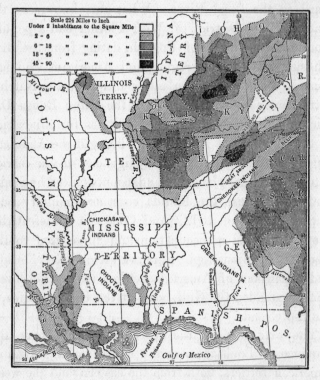

MAP OF DISTRIBUTION OF POPULATION IN 1810

from the Mississippi, on the Bayou Teche for fifty miles
below Opelousas, and on several bayous west of the
Atchafalaya in the same neighborhood. The French
occupied the Red River continuously up to Alexandria,
less compactly as far as Natchitoches, while there were
some scattered dwellings higher up the stream as also

upon its northern tributary, the Washita, above the present town of Monroe.[20] The first comers had therefore selected the rich alluvial bottoms, generally within fifty miles of the Mississippi, which formed the chief highway of the country.

In the old Florida parishes on the east bank the Americans predominated. They were to be found elsewhere chiefly in New Orleans and certain coast towns, though their plantations were scattered sparsely east of the Washita and north of the Red, just opposite the dense American center in the old Natchez District. The French occupation of the present state of Louisiana had progressed from the coast inland, and therefore had fixed upon the lower courses of the rivers. The American immigrants were later comers, and advanced from the interior of the continent; hence they located inland on the upper courses of the streams, excepting those merchants who were attracted to the seaports.

As the wild lands in the old West were taken up, and steam navigation on the rivers facilitated expansion, population poured across the Mississippi till in 1840 it occupied a broad band up and down that stream, defined on the west by the ninety-fifth meridian and the Sabine River, and reaching from Iowa to the Gulf.

Thus the purchase of Louisiana was abundantly justified, if only on the ground of American capacity for expansion; but the acquisition of so vast a territory was bitterly opposed in the Middle States and New England, the natural habitat of the provincial point of view. The reasons advanced were the preponderance of the South and West in the future councils of the nation, as this large area should be carved up into states; and the possible formation of an independent confederacy, embracing the whole Mississippi basin and thus curtailing the extent and strength of the original Union.[21] It was argued that the course of the river would divert western interests from the East to the South. The same fear had been set forth by the East at the time of the negotiation of the Jay treaty as a reason for accepting the occlusion of the Mississippi as a means of counteracting the separatist tendency of the transmontane settlements.

Fortunately the narrow policy of the New England statesmen secured no adequate following. The acquisition of Louisiana destined the United States for a continental power. The Republic was born synchronously with the idea of national empire and large territorial dominion. " Young countries know no exclusion from the sea; they are wont rather to grow from the sea landwards and if possible onwards till they reach the sea again." [22] All the countries of North America, except little San Salvador, have a foothold on either ocean, sometimes the very smallness of this foothold indicating its importance.

In South America physiographical conditions have been adverse to such sea-to-sea expansion, except at the northern and southern extremities of the continent where the mighty rampart of the Andes sinks into the ocean. Colombia has a broad base on the Pacific and the Caribbean Sea ; and where the Straits of Magellan cut through the continent, the bare rim of the mainland is held by aggressive Chile to the detriment of

Argentina. The wall-like barrier of the Andes, rising from no gently sloping plateau as do the Rockies, presented a most formidable obstacle to expansion from the east, while the tropical climate of the greater part of South America served to emphasize the indolent temperaments of the Latin half-breed population. Though the upland valleys were early occupied from the western coast, the mountain district of the Cordilleras could not soon produce a pressure of population sufficient to mount the snowy ranges and overflow into the great river valleys on the Atlantic side. But recently the Pacific states are reaching out towards the east, and the number of boundary disputes on the Atlantic slope of the Andes is significant.

Russia, the youngest country of Europe to develop under geographical conditions of continental expansion, from a small inland area has pushed her frontiers to the Baltic, the Black Sea, and the Pacific, where she is constantly lengthening her base; she promises in the future to reach the Indian Ocean through the valley of the Euphrates and the Persian Gulf, or through the Gates of Herat and Afghanistan.

Four months before the Louisiana Purchase was consummated, the Lewis and Clark expedition was planned to prepare the way for American expansion to the Pacific by discovering the best overland route for commerce. Jefferson wished to exploit the advantages of the great system of river communication across the continent; he realized the unifying power of the Missouri in creating an interest for the United States in the far West. "The river Missouri and the tribes inhabiting it are not as well known as is rendered desirable by their connexion with the Mississippi, and consequently with us." [23]   He complained that the furs and peltries of the remote Northwest were carried off by the British traders through Canada by a far less perfect system of inland navigation which was interrupted by innumerable portages and shut up by ice much of the year; whereas these products of the northwestern forest ought naturally to reach the Atlantic through American waterways.   Lewis was instructed, therefore, to explore the Missouri carefully, and follow any of its principal streams which might afford the readiest communication across the mountains with the Columbia and the Pacific Ocean.

The Mississippi valley was destined to become the core of the nation as it was of the continent.   Its fertile soil would support a dense population, and its cheap waterways were to prove of inestimable value for a young, agricultural people.   The acquisition of the new West prolonged greatly the most distinctive feature of American anthropo-geographic conditions — the abundance of free land.[24]   A nation is influenced not only by the topography, but by the size of its territory. The presence of the new West reacted most wholesomely upon the East and the old West: the stimulating effect of inexhaustible opportunity never allowed American energy to abate, and the democratic spirit of the ever youthful frontier fostered the spirit of democracy and youth in the whole nation.   The abundance of unoccupied land beyond the Mississippi afforded a solution of the Indian question, and prevented it from ever reaching an acute stage. Some of the Indian purchases were doubtless delayed because of the new outlet for the white population; but when made, were simplified by the transfer of the tribes to new hunting-fields beyond the Mississippi. Thus the western half of the great valley brought more room to the eastern half.

NOTES TO CHAPTER VI

1. Ratzel, Anthropogeographie, vol. i. pp. 349–352, 2d edition.   1899.
2. W. Z. Ripley, The Races of Europe, maps, pp. 143, 147.   1899.
3. Ibid. p. 222, map.
4. Ibid. pp. 403, 429, maps.
5. Ibid. p. 284, map.
6. Justin Winsor, The Westward Movement, p. 372.
7. Ibid. p. 32.
8. Monette, History of the Valley of the Mississippi, vol. i. p. 545. 1846.
9. C. B. Hartley, Life of Daniel Boone, pp. 308–324. Phila. 1865.
10. E. Coues, History of the Lewis and Clark Expedition, vol. i. p. 7.
11. Roosevelt, Winning of the West, vol. iv. p. 266.
12. Monette, History of the Valley of the Mississippi, vol. ii. pp. 541, 547. 1846.
13. P. L. Ford, Writings of Thomas Jefferson, vol. viii.   Confidential Message to Congress on the Expedition to the Pacific, Jan. 18, 1803.
14. Ibid.   Letter to John Bacon, April 30, 1803.
15. Ibid.   Letter to Robert Livingston, April 18, 1803.
16. Roosevelt, Winning of the West, vol. iv. p. 207.
17. Albert Bushnell Hart, Foundation of American Foreign Policy, p. 97.   1901.
18. Monette, History of the Valley of the Mississippi, vol. ii. p. 461. 1846.
19. Ibid. vol. i. pp. 83–85.
20. Ibid. vol. ii. p. 516.
21. P. L. Ford, Writings of Thomas Jefferson, vol. viii.   Letter to J. C. Breckenridge, April 12, 1803.
22. Ratzel, Politische Geographie der Vereinigten Staaten, p. 1.   1893.
23. P. L. Ford, Writings of Thomas Jefferson, vol. viii.   Confidential Message to Congress, Jan. 18, 1803.
24. F. J. Turner, Significance of the Frontier in American History, Report of American Historical Association, 1893.

CHAPTER VII

GEOGRAPHY OF THE ATLANTIC COAST IN RELATION TO THE DEVELOPMENT OF AMERICAN SEA POWER.

WHILE western settlement was spreading up the lower reaches of the Missouri, Arkansas, and Washita in 1810, the center of population for the whole country lay at the eastern foot of the Blue Ridge, forty miles northwest of Washington and the tidal floods of Chesapeake Bay. Even as late as 1830, the star which marks the westward course of population on the census maps had not passed beyond the head waters of the Potomac; it still lingered with the waters draining into the Atlantic, thus showing the ocean's power of attraction, even after the nation by the acquisition of Louisiana had committed itself to continental dominion. The influence of the Appalachian barrier, as we have seen, was to limit the British colonies to the coastal plain, and in certain regions to stimulate maritime development because of the limited supply of easily exploited natural resources on the narrow shelf of the continent. When fur-fields and corn-fields gave out, enterprising spirits sought a new field of activity on the sea.

The attraction of the abundant lands in the Louisiana country was not felt immediately. Enterprise and capital do not change their direction so rapidly. The decade of 1830 and 1840 witnessed a decline in the American merchant marine and a widespread movement of population which pushed the western frontier to the ninety-fifth meridian, where it lingered for many years. Then our development became continental as opposed to maritime or extra-continental. "In the present century we see the effects arising from the opening of the Mississippi Valley to our Atlantic coast peoples in the gradual decadence of our shipping interests. . . . Although there have been many influences at work in the diversion of our people from maritime life, it seems on the whole that the most important cause is to be found in the way opened to enterprising people through the ready access which this century has given to the central fields of the continent. This phase over, we may expect reversion to maritime life on the Atlantic coast." [1]

Through all the colonial period and for the first four decades of the Republic, the United States was dominated by the ocean. Geographical location and a variety of geographical conditions determined this. The original Thirteen Colonies were a long cordon of settlements along the Atlantic, stretching from the St. Croix River in the north to the St. Mary's in the south; they had length but not breadth, and this length was washed by the tides of the ocean. Each colony had ample contact with the sea, and developed along the lines of its sea connection. Travel from colony to colony by land crossed the watercourses at right angles; only in exceptional cases could it make use of these natural highways of a young community. Inter-colonial travel and traffic therefore went by sea, and the coastwise trade became a great artery through which circulated the common life of the nascent people, contributing to that solidarity which made their strength.

Of harbors there were plenty, from one end of the coast to the other. The Atlantic shore of the United States is characterized by multiplicity of small indentations and paucity of large ones. In this respect it presents a contrast to Europe. Chesapeake Bay and Long Island Sound are the only partially inclosed bodies of water considerable in size, and even they run parallel to the coast and so contribute comparatively little to opening up the back country. This office was performed by the Atlantic rivers with their estuaries, which sufficed to carry maritime influences well into the heart of the narrow zone of settlement.

A paucity of large inlets means a paucity of peninsulas. In the domain of the United States we can count only the Delaware-Maryland arm of land, southern New Jersey, the southeastern projection of Massachusetts, and at a later date Florida, which, however, owing to its poverty of harbors, has contributed little to the maritime development of the country. The United States has also few segregated areas in the form of islands; it lies between the great island regions of the Atlantic, between Newfoundland, Cape Breton, Prince Edward, and Anticosti to the north, and the vast loop of the Antilles to the south. Thus the continental location of the United States is emphasized.[2] Such islands as it has are close inshore, a part of the general contour of the country. They are fragments of the mainland cut off by glacial action and subsidence, as those along the fiord regions of Maine ; or long bars of deposit which inclose the coast with few interruptions from Sandy Hook to the Rio Grande.   Only Long Island by its size and Nantucket by its comparative remoteness have achieved an independent or individual existence.   The presence of foreign powers in all the islands to the north and south and even those off the coast, as the Bermudas, Bahamas, and Antilles, has been a distinct factor in the naval and military history of the United States, demonstrating the disadvantages of a purely continental location.   But now that the nation is alive to the value of such oceanic outposts and is securing them where she can, much of this will be changed.

Length of coast-line and abundance of harbors united with a third geographical condition, namely, location, to make the Americans from the start a seafaring people.   They held a central position on the rim of the populous and progressive North Atlantic basin, and in this position New England occupied the choice place. Trade followed the great sweep of the Gulf Stream, the North Atlantic Drift, and the returning Equatorial Current past the shores of Newfoundland, Great Britain, Spain, Africa, the Canary Islands, West Indies, and up the coast of the United States.   This path the American schooners followed, helped onward by the currents and their causing winds, visiting in turn all the markets passed and completing with profit the great circuit of trade.   Not England itself had a more advantageous position for the commerce of the North Atlantic, if we except her proximity to the isolated maritime field of the Baltic.   Located in the temperate zone, the colonies, as later the states, were under the necessity of importing tropical products from the West Indies, among whose islands American vessels threaded their way, carrying in exchange the substantial foodstuffs of the North. As a new country with an abundance of surplus products from forest, field, and shore, it demanded the manufactured wares of Europe, while that continent in turn clamored for the raw materials of America. To both these important markets colonial America held a near and central position. These are the large features of her geographical location. A more detailed analysis of early American exchanges will reveal other advantages which can be attributed in great part to conditions of climate, soil, and physical feature obtaining in the states themselves, or in the markets which their merchantmen sought along the current paths of the pathless ocean.)

Within the area of the Thirteen Colonies there were differences of natural condition sufficient to diversify their activities and their products, to make land interests predominate in one section and maritime in the other, and that in spite of the fact that both regions had ample contact with the sea. The geographical environment of a people consists of all the natural conditions to which they are subjected, not merely a part. A coast region shows always the interplay of the forces of land and of sea. According to the fertility and extent of arable soil, or to the abundance and availability of harbors combined with an unproductive or limited back country, now the agricultural phase of development, now the maritime prevails, following the law of the resolution of forces. A one-sided view of environment in the study of geography is disastrous, as shown by the conclusions of a recent writer, who says: "The evidence of history is, however, strongly against the assumption that geographical facilities alone will make a people maritime. The Phœnicians and the Jews dwelt side by side on the same coast, but they made very different uses of their opportunities. The Phœnicians filled the Mediterranean with their commerce. . . . The Jews never developed any aptitude for the sea." [3]

These two peoples made a different use of their opportunities because their opportunities were essentially different, both as to the land and as to their respective coasts. The Jews inhabited a broad band of fertile country stretching westward from the Syrian desert, but reaching the Mediterranean only southward from the promontory of Mt. Carmel; here the coast is one smooth, unbroken stretch, even the tiny estuaries of its streams being filled up by the current-borne sand swept northward from the Nile delta. The coast afforded no harbors, while the back country with its varied, abundant crops, rich orchards, but scanty forests, offered every inducement to an agricultural life. The Phœnicians occupied a narrow hem of land between the solid range of the Lebanon Mountains and the sea. The port of Sidon was only fifteen miles distant from the summit of the range (4000 feet). Expansion to the eastward was practically impossible; only where the Leontes River had carved its way through the mountain barrier to the sea did the Phœnicians stretch their dominion through its gap to the inner range of Mt. Hermon. The coast which they held north of Mt. Carmel, in contrast to that of the Jews, was somewhat broken and irregular from the fact that the roots of the mountains here intruded upon the sea. An occasional harbor, an inshore islet making a protected roadstead, and outlying Cyprus, invited to maritime life. On the slopes of Lebanon, which condensed the moisture brought in by the prevailing southwestern winds, grew the magnificent cedars which furnished the material for the Phœnician fleets. But the steep hillsides would grow little else. Hence geographical conditions both of the sea and of the land combined to make a maritime race of the Phœnicians, an inland people of the Jews.[4]

Much such a contrast was presented by the Atlantic shore of America and with much the same result. In New England the deeply embayed coast, the narrowness of the lowland belt, and the glaciated soil were all geographic factors operating to develop maritime life. In New York and Raritan bays similar conditions, combined with favorable routes of communication with the interior, produced a region of natural seaports with the attendant nautical activity. Philadelphia near the mouth of the Delaware River and Baltimore commanding the Susquehanna outlet became the ports for eastern Pennsylvania. But southward from this point, in spite of bays, sounds, inlets, and deep river mouths, the sea loses its ascendency in controlling the activities of the people, and the extensive, fertile soil of the southern states determines their agricultural development. By length of coast-line the southern colonies were kept in contact with oversea influences, but the profits of tobacco and rice crops withheld their investments from the carrying-trade, which they therefore left to others, notably New

England. Even to-day the southern states own scarcely more than a tenth of the ships under our flag.

The great navigators of the world have always been those nations living in small, half-barren coast countries, or in larger countries with only a limited fertile area, but lying near to larger fruitful regions for whose products they could play the part of middleman, or exchange the harvest of their seas. Such peoples were the Phœnicians, Carthaginians, Greeks, Latins, English, the burghers of the Hanse towns, and the Norwegians. The rocky, indented coast of New England produced the great maritime carriers of the New World, with whom even British merchantmen found it difficult to compete. The ports of Massachusetts Bay are four degrees of longitude nearer to England than is Philadelphia, nine degrees nearer than Charleston, South Carolina; therefore they had distinct advantages in the trade with the island kingdom which, by reason of its limited area and denuded forest lands, made a steady demand for the raw products of America. Especially it needed masts, spars, and lumber for the royal navy from the northern forests of New England. These materials the colonists also worked up assiduously for their own marine, so that shipbuilding for home uses and exportation became one of the earliest and most important industries of the northeast coast.

The geographical distribution of early shipbuilding ports in New England shows the dispersion of maritime interests along the whole coast, as opposed to the concentration at a few favored points in later times; it tells of abundant forests near all the shipyards, of the splash of waterfall and the whir of sawmill almost mingling with the roar of the surf on the shore. In the period from 1620 to 1750 we find vessels being launched at numerous towns along the coast from Cape Ann to the Connecticut River, — at Gloucester, Salem, Marblehead, Medford, Dorchester, Boston, Plymouth, Portsmouth and Newport in Rhode Island, Westerly on the estuary of the Pawcatuck, and New London.

But after 1750 the lumber supply near the older ports was exhausted, and the shipbuilding industry began to concentrate along the shores of northeastern Massachusetts, New Hampshire, and the nearer coast of Maine. Wells, Maine, built a schooner of eighty-eight tons in 1767, but the greater center of shipbuilding activity was on the Piscataqua River, where vessels were turned out at the rate of two hundred a year. The abundant forests furnished the lumber; the cordage, anchors, and canvas to equip the vessels came from England, though Portsmouth started up three ropewalks at this time. Naval stores, such as tar, pitch, and turpentine, came from the northern woods or the forests of North Carolina. As the Piscataqua began to need all its lumber resources for the construction of ships, the center of export of masts and spars moved eastward from Portsmouth to Portland, Maine, evidencing the relationship between forests and commerce.[5]

Reading the early commercial history of New England, one seems never to get away from the sound of the shipbuilder's hammer and the rush of the launching vessel as she " schoons " on the waves. Massachusetts employed a fleet of two hundred vessels as early as 1709, and sold on an average one hundred a year to merchants of London and elsewhere.[6] Shipbuilding was inspected by the government, such was the pride in the industry, and builders were granted special privileges in several ports. Other lines of manufactures which started up before 1650 remained undeveloped because this one great industry absorbed the energies and capital of the population. American ships were the cheapest and best afloat. Philadelphia, which was advantageously situated near the Appalachian forests and also near the sea, turned out the best-finished vessels, but New England's were the fleetest and staunchest.

Such an important field of activity was offered by the American merchant service that it attracted a superior class of men both for officers and crews. A vessel which started out with the highly composite cargo of an American ship, to visit the markets on the great circle of trade from Newfoundland around to Barbadoes, required a master who understood the possibility of every market both for selling and buying, and was alert to pick up a bit of freight business between the foreign ports which he might happen to pass. For this reason the *personnel* of American seamen was far superior to that of the British in the eighteenth and early part of the nineteenth century. The consequence was that American vessels were the favorite carriers on the ocean. They were the fastest, safest, took better care of their goods, loaded and unloaded most quickly. This reduced for them the rate of insurance, and enabled them to select the most valuable freight at the highest profit.[7]

The shipbuilding industry of New England was stimulated especially by the proximity of the extensive fishing-grounds off the northeast coast of North America, where the submarine plateau of the Grand Banks pastured the finny herds of the ocean. Fisheries have always been the original inducement for men to go to sea, and they continue to-day to be the training-school of seamen. Boston alone in 1664 had three hundred boats fishing in the waters about Cape Sable; and there were fifteen hundred fishermen casting in their nets off the Isles of Shoals.[8] Cod became the staple of New England exports. Salted and packed, it found a ready and extensive market. The choicest fish were sent to the Catholic countries of southern Europe where the regular fast-days occasioned a steady demand, just as they had done in the fifteenth century for the salt herring of the great Hanse fisheries.[9] The medium quality was kept for home consumption, and the refuse fish was carried in great quantities to feed the slaves on the sugar plantations of the West Indies. This trade was carried on regularly with the British holdings in the Antilles, and illicitly but extensively with the French and Spanish colonies. The limited area of these islands and the more profitable use of their soil for a few semi-tropical products necessitated the importation from the American colonies of foodstuffs, which presupposed larger areas for agriculture and pasturage, as well as proximity to the northern fish. At the outbreak of the Revolution, the interruption of commercial dealings with the West Indies almost annihilated the fisheries and likewise the slaves, for thousands of these unfortunates died of starvation in a few years.

But cod and mackerel were not the only prizes that lured the American fisherman upon the water. In the early colonial days the right or baleen whale frequented these coasts, and a spout offshore was the signal for the launching of boats for the attack. In case a dead whale drifted to land, riparian rights, defined by law, decided the ownership of the prize. But soon the drift whales decreased, and in consequence sloops and other small craft began to seek their prey, especially the sperm whale, in the open ocean, and a regular industry developed. This seems to have been concentrated chiefly on the southeastern coast of New England. The leaders were Nantucket, Long Island, and New Bedford; Providence, Warren, and Newport in Rhode Island, Dartmouth and the Cape Cod district participated. Boston took no very active part, though it was the chief port for the commerce in the products of the whale.

Deep-sea whaling demanded larger vessels, both for safety and to hold the big cargo of bone and oil. These ranged first from the Bahama Islands northward to Baffin Bay and Davis Strait, then across to the Azores and the Guinea coast. New England soon led the world in this industry; and when the whale was exterminated in the North and South Atlantic, her vessels followed it to the Pacific and Arctic. Whalebone and oil became important items in domestic and foreign exchanges. Of five thousand tons of whale-oil imported into London in 1763, nearly three fifths belonged to American owners, while forty tons of bone were

deposited in the same market in 1761 and 1762.[10]
Thus ships, lumber, fish, whale-oil and bone, New
England's contributions to the world's exchanges, were
intimately connected with her geographical location and
natural features of coast-line, climate, and soil.

(Geographical conditions, as we have shown above,
determined that inter-colonial communication should be
largely by sea. This prepared the way for a peculiarly
active coasting-trade, especially as every district of the
Atlantic slope was highly differentiated as to its products in consequence of varied natural conditions. The
geographical control in the activities of New England
and the southern states has been considered. New York
and Philadelphia, by reason of their extensive interior
connections through the Hudson, the Delaware, and the
Susquehanna, were the natural ports for the fur trade.
New York, Pennsylvania, and northern Maryland were
the chief wheat-producing countries. New England,
with the early exhaustion of her limited farm-land, soon
(1745) found it necessary to import her wheat,[11] and
the southern states found it more profitable economy
to get their cereals from the North. South Carolina
utilized her swamps for rice, and North Carolina, which
long held the place of a frontier country to the older
Virginia, contributed to commerce the products of
a frontier region, — lumber, naval stores, cattle, and
leather.)

The most active agents in the early coasting-trade
were the Dutch of New Amsterdam, whose central
position in the angle of the continent gave them peculiar advantages; but they soon found active competitors

in the New Englanders, as soon as these later comers began to adapt themselves to their environment. The absence of great inlets along the western shore of the Atlantic had the effect that the American coastwise commerce was not thalassic but oceanic. It made wide sweeps, and in truth was intimately associated with the foreign trade. Just before the outbreak of the Revolution Rhode Island alone had three hundred and fifty-two vessels coasting from Newfoundland to Georgia.[12] The small coasting-craft did a continuous drop-in business from port to port. The southern states wanted fish from the Grand Banks, European wares and wines from Boston, West Indian molasses, sugar, and rum from the New Haven docks, and wheat from New York and Pennsylvania; while tar from Albemarle Sound and tobacco from the Chesapeake in turn went up to fortify the hulls of the ships and the souls of the skippers off the Grand Banks. Often a homebound vessel from a tour in the Caribbean Sea would stop at Norfolk and exchange a part of its semi-tropical cargo for tobacco. Thus the coastwise trade shaded off into the foreign trade with the near-by Antilles.

But the commerce which was most remunerative and most regular was that with the West Indies. The islands demanded things which the colonies could give them or bring them, and they gave in exchange products which commanded a ready market. Besides fish, salted meats, and flour, they wanted horses in large numbers for the cane-crushing mills of Barbadoes and other sugar districts, barrel and pipe staves for their sugar and molasses, boards and houses ready framed to be set up, and lastly they wanted slaves. The slave-trade, particularly, was profitable and delightfully simple in its reciprocity aspects. Moreover, it stimulated the industrial development of New England by introducing the manufacture of rum in enormous quantities. Made of inferior molasses which the Indies themselves furnished in vast quantities, rum was distilled chiefly at Boston and in the twenty-two still-houses of Newport, and then went to buy the kidnapped negroes on the Guinea coast and introduce the residual inhabitants to the advantages of civilization. The slaves, who were rescued from the debauching effects of the rum, closely and carefully packed as it behooves valuable merchandise, made a short voyage to the Antilles, were distributed among the British, French, and Spanish plantations, or were brought further north to supply the slave-markets of the southern states, without contaminating the soil of New England, or by their presence disturbing the placid conscience of even such a liberty-loving trader as Peter Faneuil.

The chief places interested in the West Indian trade were Ipswich, Salem, Boston, Newport, New London, New Haven, Windsor, and New York. We notice that Long Island Sound points, which were slow in participating in colonial maritime activity, became important in the West Indian trade. The wealth of New Haven was based largely on its commerce with Barbadoes. New England vessels visited with little discrimination British, Spanish, French, and Dutch holdings in the Caribbean region; for the Navigation Laws were too carelessly enforced to limit the commercial field of the enterprising Yankee trader.   Outbound vessels from the American coast generally carried an extra supply of staves and a number of coopers, who were busied setting up barrels and hogsheads to hold the return cargo of sugar and molasses.   The same vessels brought in also indigo, cotton, mahogany, dyewoods, silver, and Spanish iron, the last two being items of Spain's trade with her own possessions.   New England, always in need of salt for her fish industry, at one time tried importing this article from the Tortugas of the Gulf, but it proved an unsatisfactory variety for preserving fish, and she continued thereafter to rely upon the Mediterranean supply, which was shipped chiefly at Malaga.

As England's commercial policy permitted the colonies to trade with all European countries south of Cape Finisterre in such commodities as she herself did not produce, American merchantmen did a thriving business with Spain, Portugal, and, on the return voyage, with the Madeira and Canary Islands.   These countries being Catholic wanted fish, and being great wine-producers wanted staves, both of which the colonies gladly furnished for salt, wine, oil, and fruits. Once in these regions, it was a simple matter to run down the African coast for gold-dust, ivory, and slaves.

Geographical conditions determined New England as the great maritime region of America.   New York's limited though valuable coast, and the retardation of her back country by the proximity of the hostile French, and later by the retention of British posts on the Lakes, prevented her early maritime development. In the middle of the eighteenth century, the entrances and clearances in the ports of Philadelphia and New York combined were just equal to those of Boston alone. In 1731, when Pennsylvania owned 6000 tons of shipping, Massachusetts had nearly 38,000 tons, one half of which was employed in European trade. In her early history, the port of New York had to share honors with Perth Amboy, which attracted trade on account of its nearness to the sea and its superior advantages for avoiding the import and excise taxes;[13] but with the expansion of population along the shores of Ontario and Erie, the activity along this frontier in the War of 1812, and the consequent construction of the Erie Canal, the port of New York attained that preëminence marked out for it by its position before the " eastern gateway of the United States."

A merchant marine becomes in time of war an important factor in national defense and aggression. It furnishes seamen, transports, its vessels are converted into armed cruisers, or are fitted out with letters of marque to prey upon the enemy's commerce. Colonial privateers did effective service in the Spanish and French wars, for it was easy to intercept French supply ships or merchantmen in the Gulf of St. Lawrence, or Spanish vessels on their way to the Antilles. In the Revolution, the supremacy of English sea power left any regular naval warfare out of the question for the young Republic. However, the great number of her merchant ships and seamen, whose occupation was for the time destroyed; their familiarity with the secluded ports and island hiding-places of the coast; moreover, England's necessity of exporting her supplies in this

oversea war, while she held a naval base on the mainland
only at two points, New York and Newport, all combined to make a situation in which privateering could
be most effective, while affording to the merchant marine
means of partially recouping their losses.   Therefore we
find that Boston had three hundred and sixty-five vessels
commissioned for this semi-piratical service, Salem about
a hundred and eighty, Rhode Island some two hundred,
though right under the nose of the British guns at
Newport, — while other New England ports frequently
sent out vessels.   That they went to some purpose is
evidenced by the fact that in 1776 the rate of marine
insurance between England and the West Indies rose to
twenty-three per cent.   The vessels which the national
government put into the service were most effective as
cruisers on the high seas, while the colonial cruisers,
with the privateers, were employed in the same way inshore.   The prize cargoes of military and naval supplies
which were thus brought in did much to relieve the
needs of the colonies, which, in consequence of their
previous dependence upon England, were almost destitute of the simplest means for carrying on the struggle.
From Cape Breton to Martinique these gadflies of the
ocean stung, preyed upon, and almost stampeded English commerce.

Just as the achievement of independence was the
signal for territorial expansion on the part of the Americans, so it marked the beginning of a pronounced
maritime expansion.   England was to be no longer the
middleman in the trade with the Orient.   American merchantmen, encouraged by discriminating duties in their

favor, found their way to Canton and the East Indies, stopping *en route* in the Indian Ocean to trade at Bourbon and Mauritius, which had been opened by France to American commerce. A Boston ship, in 1788, following in the track of the whalers in the Pacific, began trading with the Indians of the northwest coast of America for furs, which it carried across to Canton and exchanged for Chinese products. In the succeeding years New England vessels were to be seen constantly in the Spanish ports of California, where they outnumbered all other foreign merchantmen, and first drew the attention of the national government to the need of a foothold on the Pacific.

The exclusion of English competition by the Navigation Laws and superior facilities for shipbuilding enabled the states to exploit all the advantages of their location and natural equipment for maritime ascendency. While in the period from 1789 to 1793 their proportion of the carrying-trade of the country rose from twenty-five per cent. to seventy-nine per cent., by 1810 they had absorbed eighty-nine per cent. Interests so important and operating in such a wide field required the protection which only a navy could give. Depredations on American commerce by French vessels in the West Indies in 1793 was the occasion for the construction of the first war-ships by the new Union and for the reprisals which taught the French manners. Similar outrages by the corsairs of the Barbary States, just outside the Strait of Gibraltar, gave the youthful navy, in 1803, that experience in conducting a campaign which prepared it for the more serious contest of 1812.

NOTES TO CHAPTER VII

1. Shaler, Nature and Man in America, p. 200.
2. Ratzel, Politische Geographie der Vereinigten Staaten, p. 29.   1893.
3. George, The Relations of Geography and History, p. 76. Oxford, 1901.
4. George Adam Smith, Historical Geography of the Holy Land, pp. 76–85 and 127–135.   1897.
5. Weeden, Economic and Social History of New England, vol. ii. pp. 765, 766.   1899.
6. Ibid. vol. i. p. 366.
7. Shaler, The United States, vol. i. pp. 551, 552.
8. Weeden, Economic and Social History of New England, vol. i. p. 245.
9. Schaefer, Die Hansestädte und König Waldemar von Dänemark, p. 256. 1879.
10. Weeden, Economic and Social History of New England, vol. ii. p. 746.
11. Ibid. vol. ii. p. 507.
12. Ibid. vol. ii. p. 761.
13. Broadhead, History of the State of New York, vol. ii. pp. 392 and 460.

CHAPTER VIII

In the War of 1812, even more than in the Revolution, the United States felt the advantages of its remote geographical location and large area. The three million people who had fought the War of Independence had increased in 1810 to over seven millions, a consequence of the abundant land and resources of a new country, its encouragement of natural increase, and attraction to immigrants. The line of inhabited coast was only little longer than in 1780, but the land frontier of continuous settlement slanted southwest from Lake Champlain to the mouth of the Ohio, and from that point southeast to the St. Mary's River, which marked the northern border of Spanish Florida. Outside of this 2900 mile frontier were isolated areas of settlement, every one of which, except the Mississippi group between the Ohio and the Missouri, figured in the War of 1812, — Detroit, Mackinaw, Prairie du Chien, Fort Dearborn (Chicago), New Orleans, and Forts Mims and Bowyer which guarded the approaches to the Tombigbee settlements.

From the geographical standpoint, the distinguishing characteristic of this war lay in the fact that it was limited strictly to the frontiers, both land and marine.

Only in the attacks on Washington, Baltimore, and New Orleans, which, by reason of their tidewater location, can be set down as coast towns, did invasion penetrate more than a few miles from the sea; while on the land frontier, where there was no waterway to aid, the British hesitated to advance even this far from their base upon the border. It was therefore a peripheral war, and left the mass of the country undisturbed. This could have been the case only in a country of large area and isolated location, or in a land like England, with only marine frontiers defended by a strong navy. Thus even the taking of Washington, the capital, lost its military significance, because it only scratched the skin of the country and left the great body unimpaired. The fall of Pretoria in a ten-fold larger Transvaal would not materially have altered the outlook of the Boer War, just as the taking of Moscow in 1812 shrank to insignificant proportions against the background of Russia's vast area.

In view of the maritime development of the United States at this time, the impressment of American seamen into British service had to be checked. Six thousand Americans on English naval vessels tells how the English had improved their opportunities. The war was undertaken, therefore, to vindicate the rights chiefly of the maritime states, and yet it was just these states who were opposed to the war. Hostilities meant the interruption of navigation and the destruction of their commerce; but they preferred to lose their mariners rather than their money. Hence the sectionalism of parties in their attitude towards this war was clearly drawn.

In the opposition was found all of New England, with
the exception of the coastless state of Vermont, a large
part of New York, and the majority of New Jersey and
Delaware. The South and West supported the war,
though Maryland, showing the conflict of her land and
sea interests, cast three votes against it.[1] The leader
in the opposition, as was to be expected, was Massachusetts, moved by her dominant maritime interests.

The war was probably distasteful to New York because she would suffer not only in her port, but also
in her northern frontier, which, together with that of
Vermont and the northeastern corner of Ohio, constituted the only settled area of the United States contiguous with the Canadian border. This was the old
danger-line from Champlain to Presque Isle (Erie)
which had been made conspicuous in the French wars
and in the Revolution; and now with the advance of
population it extended westward to Detroit. Moreover,
in contrast to the natural indifference of French Canada
in this conflict, the Americans felt the animosity of
their neighbors in the adjoining province of Ontario.
These were loyalists who had withdrawn from the United
States after the Revolution. Opposed to mingling with
the French element in the older eastern part of Canada,
like all late comers in America, they were located in
the west, and were granted tracts of land by the British government in the remote district north of Lake
Ontario, where they formed the nucleus of the later
political division of Upper Canada.[2]

It was along this line that hostilities began, though
war had been declared because of maritime outrages.

But the youth and weakness of the American navy, and the strength of the British in our waters, with their naval stations at Halifax, the Bermudas, Santa Lucia, Barbadoes, and Jamaica, whence they could blockade our coasts, necessitated the opening of the campaign with an attack on Canada. But here, again, geographical conditions determined that the war should be in large part naval; fleet and infantry were to combine. Lakes Ontario, Erie, St. Clair, and Champlain were the scene of these water operations; and the two short, detached lines of the Niagara and Detroit rivers, where the land frontiers converged, became strategic areas of continued military activity where encounters were hand to hand.

Along this inland frontier the British had the advantage. The St. Lawrence afforded them a protected line of communication with the naval and military stores of Canada and England, and hence brought the frontier nearer to the treasury and troops of England than were the Americans to the Hudson. The Mohawk route with its imperfect and interrupted navigation, though it had served well for the birch canoes and furry cargoes of the colonial wilderness, was ill adapted to the transportation of heavy guns. Moreover, the outlet of this route was the mouth of the Oswego River, and supplies for the naval base at Sacket Harbor at the easternmost end of Lake Ontario had to run the gauntlet of British ships which issued from their station at Kingston near by; while Kingston, by its position at the effluence of the St. Lawrence, secured the British line of communication on that river. For both sides the control of these inland waters was of the utmost importance, because upon it depended the supplies of all the Niagara and Detroit River posts.

The land position of the British, too, had some geographical advantages. Lake Ontario laps over the end of Lake Erie in such a way as to make a small rectangular peninsula of the British territory along the Niagara River; and this formation is duplicated at the western extremity of Ontario Province along the Detroit River by the same relative position of Lake Erie and Lake St. Clair. The position of the British, flanked by the Lakes, was peculiarly protected along these two rivers, which, by their shortness, isolation, controlling positions, and their exceptional character as the only settled land frontiers west of the St. Lawrence, were immediately endowed with strategic importance. The British post of Fort Malden particularly, by its location on an angle of land at the mouth of the Detroit River, was able to cut off reinforcements moving to Detroit by land and supplies coming by water from the American settlements at the eastern end of Lake Erie.

The long arrow-shaped wedge of Upper Canada brought the British at Malden far into American possessions, enabled them to cut off the upper lakes from Erie, and gave them an administrative center on the northwestern frontier for stirring up the Indians all the way from the Wisconsin River to Sandusky Bay. Detroit was the post whence the efforts of the British might be frustrated: hence its strategic importance and the early movement of American forces towards this point. The capture of the sole American boat on Lake

Erie as it was going up the Detroit River with the supplies of the troops from the Maumee, the defeat at Huron River of the reinforcements for Detroit, the capture of the post, the two battles on the Raisin River, and the advance of the British to Sandusky River, show the disadvantage of Detroit's position, remote from the frontier of continuous settlement, while British ships held the control of Lake Erie. To maintain an effective line of communication by a long detour around the western end of Lake Erie through pathless forests was almost impossible. And the British, masters of Lake Erie, were able to seize the remote post of Mackinaw, whose position was made hopeless by the interruption of its line of communication through Detroit. Indian allies seized Fort Dearborn and Prairie du Chien, though a relief party coming up the Mississippi from St. Louis endeavored to save this post at the mouth of the Wisconsin.

But while these successes were falling to the enemy, Perry was doing the impossible at Presque Isle, — converting a forest into a navy. The falls of the Niagara necessitated the construction and maintenance of separate flotillas on Lakes Ontario and Erie, thus increasing the expense and difficulty of war on the inland waters. The one American ship on Erie had been captured. The British at Fort Erie prevented supplies and a few purchased ships from coming out of the upper Niagara River to Presque Isle. Out of the green woods of Erie, therefore, from hull to mast grew Perry's fleet in 1813; there it was equipped and manned, though supplies had to be hauled over blazed tracks through the forest, and remoteness made it difficult to get seamen. But before the summer was over, this Rhode Island man with his high-seas experience wrested the control of the lake from the provincial seamen of Canada, cut off the British supplies, and cleared the way for the American advance on Malden, with the consequent retreat of the enemy from this post and Detroit, and their defeat at the Battle of the Thames.

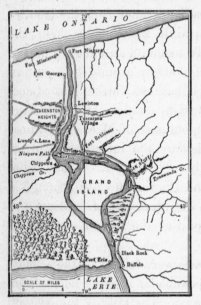

NIAGARA RIVER 1812–14

The remoteness of Detroit and the isolation of Lake Erie, due to the interruption of navigation at Niagara Falls, made this region drop out of the conflict as soon as it was retaken from the British. The really strategic frontier was the Niagara River, and here the land operations of the war were focused. Here was the battle of Queenstown Heights, the attack on Fort George, the capture of Fort Erie, the battle of Chippewa Creek and of Lundy's Lane on the Canada side; and on the American the capture of Fort Niagara and the burning of Youngstown, Lewiston, Manchester, Black Rock, and Buffalo. As the fate of the Niagara posts on both

sides depended upon the maintenance of navigation on
Lake Ontario, and as the fleets were pretty evenly
matched, neither dared risk an engagement without decided advantages of wind and position.   Too much was
at stake.   Hence warfare on Lake Ontario was color-
less because it lacked a decisive battle.   The body of
water was so small and distances so short that it was
always possible for either Americans or English to avoid
an engagement by running into a home port and taking
refuge under the land batteries.   For this reason seizure of stores in course of transportation up the Lake,
and predatory descents like those upon Oswego, Sacket
Harbor, Toronto, or Burlington Bay made up the
story of Ontario warfare.

The possible prize the British hoped to gain by successes on the Great Lakes was the future exclusive right
of navigation, by which the British could have retarded
the whole development of the Northwest : hence the concentration of military activities on Lake Ontario.   The
old Champlain route, which had figured prominently
heretofore in every war across the Canadian boundary,
by reason of its detached location played a smaller part
in the hostilities of 1812.   But when the temporary
cessation of the Napoleonic War in 1814 released
the English troops from service on the Continent,
and large reinforcements were sent to Canada for an
invasion of New York, this shortest line of communication between the St. Lawrence and the Hudson became
the line of march.

The route of the invading army from Canada was
flanked for a long distance by the Richelieu River and

Lake Champlain, so a flotilla of ships accompanied the land forces to carry supplies and at the same time lend protection. Hence both the land and water route had to be contested by the Americans. In anticipation of such an invasion, a flotilla had been stationed on Lake Champlain, but the body of water was so long and narrow that cruising was out of the question; so the vessels took up their position inshore near the land forces, which were stationed along the Saranac River to check the enemy's advance. Modern warfare is largely a science of communications and transportation. Hence the defeat of the British ships on Champlain was the signal for the retreat of the British army. Here, as on Lake Erie, a naval victory was of supreme importance to the Americans, because of the difficulty of constructing a new fleet on this isolated body of water, were the first one destroyed. Defeat would have put them in the position of the British after Perry's victory.

It was not possible for the United States to sustain any considerable damages along a land frontier which through Maine and New Hampshire was a primeval wilderness and from Vermont westward was only sparsely settled, and which, moreover, by its very shortness left only a limited area exposed to attack; but the long and densely populated sea frontier presented a vulnerable surface to the onslaughts of the British navy. From one extremity of the coast to the other, and from the beginning to the end of the war the enemy's ships made their depredations. England's great sea power enabled it to blockade all the principal harbors from

Maine to Chesapeake Bay, with the result that several
new American vessels which were building never got
to sea. This had been the case before in the Revolution. The fisheries of New England were almost broken
up and the foreign commerce of the eastern states was
totally destroyed. Stonington, Connecticut, and Lewiston on Delaware Bay were bombarded. The salt-works
on Cape Cod saved their buildings by a ransom. In
Chesapeake Bay depredations extended from Norfolk to
Havre de Grace at the mouth of the Susquehanna. The
estuaries of the Patapsco and Patuxent afforded easy
lines of approach to Baltimore and Washington, and the
Potomac carried the British fleet to Alexandria. Thus
the waterways of Chesapeake Bay and its estuaries, together with the weak-kneed resistance of the militia
which met the invading force at Bladensburg before the
city of Washington, enabled the enemy to penetrate
farther into the interior than at any other point along
the whole coast.

Further south, the coast of North Carolina was protected from naval attack by the " skaergard " of the offshore sand-reefs, for the only place to suffer was the village of Portsmouth, which occupies an exposed position
on the Ocracoke Inlet into Pamlico Sound. The South
Carolina coast was plundered by the British ships after
he defeat at New Orleans, where the enemy made their
eatest demonstration during the war. From the island
of Jamaica came the expedition, — a fleet of fifty vessels carrying twelve thousand troops. Lakes Borgne
and Ponchartrain and the numerous bayous offered
many avenues of approach to the city, whose position

was therefore even more exposed than had been that of Washington. But it was defended by troops of the backwoods and Jackson, the man of the frontier, the man of initiative and resource, the man who waited not for precedent. Still breathless from his unauthorized punitive attack upon the Spanish in Pensacola, who had aided the British in an attack upon the Americans at Fort Bowyer, he hurried back to New Orleans. He seized the cotton bales of protesting merchants to build his barricade, and afterwards paid the fine imposed for this high-handed, high-minded act. The bayous he obstructed. He set up batteries at the mouth of the Bayou Sauvage and far down the Mississippi to block approach from Lake Borgne and the Gulf. Then his men behind their cotton bales, flanked by river and marsh, showed what kind of marksmanship the military school of the backwoods produced.

It is fair to say that the only valiant work of this war was done by the infantry, regulars and volunteers, trained on the frontier, and the mariners who had learned their lesson on the high seas. Geographical environment, making the necessity and opportunity, had in both cases brought forth a sturdy product. From Newfoundland fishermen, Nantucket whalers, New England sailors, merchantmen, captains, and American shipbuilders was combined a navy that could deal some heavy blows even to the sea power of England. The British fleet was big enough to blockade every American port, but it did not know well enough the secrets of the shore to prevent the escape of sea-rovers, some belonging to the navy, but far more fitted out by private enter-

prise to prey upon British commerce. American privateers swarmed on every sea, while the navy, both men
and ships, was recruited from the merchant marine.

The sea-fights of this war, if studied merely in their
chronological sequence as presented in the ordinary
school histories, leave only a confused impression, of
which the student, young or old, retains little at all
and less that is valuable. But an analysis of the geographical distribution of these engagements reveals a
wide underlying system which explains their purpose
and brings order out of an apparent chaos.

Two important ports of the British on the American
continent were Quebec and Halifax, one their supply
station for the war along the Canadian frontier, and the
other their chief naval base in American waters. Hence
American vessels cruising to the southeast of Cape Breton Island were likely to intercept supply ships for these
two points; furthermore, they were in the path of homebound English merchant ships returning heavy laden
from Caribbean ports. But before the American cruisers
could reach this strategic position on the high seas, they
had to run the gauntlet of the enemy's ships coming
out from Halifax.

In the light of these geographical conditions, observe
some of the important naval engagements of this war.
Captain Broke, with the British vessels Shannon and
Tenedos, was instructed to watch the American coast
east of Cape Cod and prevent the escape of the four
chief American frigates from the harbor of Boston.[4]
Two of these slipped by him, but a challenge lured Lawrence in the Chesapeake to single combat, with the result

of the encounter off Cape Ann, the defeat of the Chesapeake, and the towing of the prize to Halifax. The
Constitution, which had been hanging about the entrance to the Gulf of St. Lawrence off Sable Island in
August, 1813, was victorious in a terrific fight with the
British ship-of-war Guerrière southeast of this point
(41° N. L., 55° W. L.). About halfway between Bermuda and Halifax (37° N. L., 65° W. L.) the American
sloop-of-war Wasp two months later fell in with the
Frolic, which was convoying a homeward bound fleet of
six British merchantmen of the Honduras trade. In
the conflict the Wasp was successful, but the victor and
its prize were immediately captured by the Poictiers
and taken to Bermuda.

The proximity of the British bases to the American
coast enabled [5] both regular ships and privateers to make
depredations on American commerce. Hence some of
our vessels were kept on the eastern coast to guard
against this threatening danger. The ship Enterprise,
of the United States, in the summer of 1813, captured
the Boxer from the Bay of Fundy off Penguin Point
in Maine, then cruised south and took the British pri-
vateer Mars off the coast of Florida. In the same
vicinity the Peacock from New York captured the Eper-
vier, and carried it into Savannah, the nearest American
port. Vessels from these southern ports also used to
lie in wait offshore for Jamaica convoys.

Further south, in the great tropical highway of the
trade-winds, from one side of the Atlantic to the other,
there were rich prizes to be secured in the British ships
bound for the Caribbean holdings, and also those turn-

ing southward from Cape St. Roque to round the Horn
for the East Indies. Therefore some of the fiercest
engagements of the war took place along this path of
the Indiamen. The vessel United States took the
Macedonian a few miles west of the Canary Islands.
The Hornet captured the British ship Peacock off the
coast of British Guiana, and the Southampton captured
the American Vixen among the West Indies. Thirty
miles off the northeast coast of Brazil, where the sea-
road forked to the West and the East Indies, the Java,
bound for the Orient, made a valiant resistance to the
Constitution, but was captured and then burned, because
there was no American port sufficiently near where it
might be taken for repairs. The United States suffered
in this war because of its strictly continental position
and its consequent lack of naval bases away from its
own shores. Captain Porter in the Essex made a prize
of the British packet Nocton, just northeast of Cape
St. Roque. As English influence was dominant on
the whole South American coast, and the United States
had no depot in that region, Captain Porter turned
south and rounded the Horn to levy for supplies on the
English whalers in the South Pacific, who always went
well provided, and at the same time to protect American
whalers who were at the mercy of English cruisers.
Off the coast of Chile, among the Galapagos Islands
and as far west as the Marquesas group he cruised, capturing whalers and privateers, but finally off Valparaiso
he had an engagement with the British ships Phœbe and
Cherub and was taken.

Depredations on British vessels in English waters

constituted natural reprisals. There we find the Argus making a daring cruise off the coast of England and Ireland, capturing twenty merchantmen, until she herself is captured by the British Pelican. Then her place is taken by Wasp II., which captures the Reindeer and puts into the friendly port of L'Orient on the coast of Brittany. This was the only remote region where the Americans had a friendly harbor into which they could go for supplies or repairs, or to sell their prizes. Elsewhere prizes had to be manned by the conquering ship or destroyed at sea.

Even after peace had been declared, war lingered upon the sea along the trade-wind track of the Indiamen. The Constitution captured two British vessels off the southwestern coast of Portugal, and the Hornet took the British Penguin off the coast of Brazil. During the three years of the war, fourteen hundred British vessels were taken, together with several thousand seamen and rich cargoes. Though the Americans, too, lost heavily, the spirit of maritime enterprise was kept alive and the quality of American seamanship was proven. The coastwise and foreign trade were greatly reduced in the period of hostilities, but by 1819 the reaction set in, and American vessels were doing seventy-eight per cent. of the country's carrying-trade; and from 1820 to 1830 they reached their high-water mark of ninety per cent.

After 1830, the center of population moved across the Allegheny Mountains; the development of the country became more continental, and attracted landward the activity of the people. The Erie Canal, whose necessity had been made apparent by the War of 1812, became a channel through which flowed a westward streaming tide of population, pushing the national center of gravity farther from the coast. In the South the slave power, searching for unexhausted lands which would repay cultivation under an extensive system of agriculture, was spreading over a larger interior area. Hence the ascendency of the maritime coast declined until the time when, the industrial conquest of a vast country having been accomplished, there should come an extra-continental development, seeking new markets and foreign bases of commercial activity.

NOTES TO CHAPTER VIII

1. Ingersoll, Second War of the United States with England, vol. i pp. 48–65. 1845.
2. Morris, History of Colonization, vol. ii. p. 98. 1900.
3. Ingersoll, Second War of the United States with England, vol. i. p. 198.
4. McMaster, History of the People of the United States, vol. iv. p. 91.
5. For distribution of British naval bases, see Mahan, Influence of Sea Power in History, map, p. 532.

CHAPTER IX

SPREAD OF POPULATION IN THE MISSISSIPPI VALLEY AS AFFECTED BY GEOGRAPHIC CONDITIONS

THE westward expansion of the American people has been marked by a slow advance from tidewater to "fall line," and from "fall line" across the Alleghenies; a rapid progress downstream to the Mississippi and upstream along its western tributaries to the margin of the arid belt; a leap across the Great Plains and the Rockies to the Pacific, long accepted as the outer edge of American dominion, till a faltering step was planted on the Hawaiian Islands, and a bold stride took the flag across the " world-ocean " to the Philippines.

For twenty-five years after the conclusion of the War of 1812 there was a pronounced movement of population into the Mississippi valley. This was due to the release of new forces after the cessation of hostilities; the final defeat of Tecumseh's warriors in the Northwest as allies of the British and Jackson's successful quietus upon the Creeks and Cherokees, who had long checked the expansion of Georgia and Alabama; the acquisition of the Floridas by which the Gulf States came into their own, geographically speaking; finally the forced opening up of western New York incident to the military operations on its frontier and the consequent construction of the Erie Canal. The strategic

necessity of an exclusively American waterway to Lake
Erie in the event of another war with Canada, because
of the serious difficulties and delays recently experienced
in forwarding supplies to the front, and the immediate
demand for better means of transportation through the
Mohawk thoroughfare for men going west and products
coming east, called for the union of the Hudson and
Lake Erie. Steam navigation on lake and river, well
established at this time, was developing the full useful-
ness of the interior waterways, both as avenues of immigration and means of exportation, — men coming in
one season, the products of their fields going out the
next. Moreover, a strong tide of immigration from
Europe, set in motion by the subsidence of the Napoleonic wars just at this time, found its way to the
unoccupied lands of the Mississippi valley.

All of these causes combined started a movement
towards the great central basin so strong that it went
far before its energy exhausted itself. This is the explanation of the fact that the period from 1810 to 1820
blocked out the work of expansion which the next two
decades were occupied in completing.[1] As has been
shown before, the degree of life or movement in a
people is indicated by the advance of its frontier, the
truest index to which is found in the bulges of its outer
line of settlement. In American history the noticeable fact is that these bulges have almost always been
along the course of rivers. In the early eighteenth
century these protrusions followed the Mohawk and Potomac, and later extended over the adjacent watershed,
halting along the ultramontane river sources when these

occupied fertile valleys, as in the case of the Youg
geny, Monongahela, and Holston, or passing over
more rugged plateau area to the smiling plains of the
middle courses, as in the case of the Kentucky and
Cumberland settlements.

In 1820 the frontier protrusions took the form of
long fingers, which seemed to point the line of advance
along the waterways of the country. These fingers
were longer, more slender, more eager, so to speak, south
of the Ohio and west of the Mississippi, attesting their
origin in the population of the South Atlantic states, in
whom had long been bred the spirit of expansion. One
finger crooked around the western rim of Lake Erie to
Lake St. Clair, and ten years later touched the outlet of
Lake Huron. Up the Wabash and the Kaskaskia, up
the Mississippi, almost to the mouth of the Des Moines,
up the Missouri two thirds across the present state of
that name, up the Arkansas, the Washita, and the Red,
pointed these sign-posts of western migration, while corresponding lines of settlement down the Pearl, Pasca-
goula, Tombigbee, Alabama, and Chattahoochee led the
way to the Gulf. Rivers present the lines of least
resistance to the incoming colonist, and afterwards lend
themselves to his economic needs.

Within this ragged frontier are numerous vacant
spots. These are often rough mountain regions, as in
the Adirondacks in New York, the Allegheny Plateau
in northwestern Pennsylvania, and the Cumberland
Plateau in West Virginia and East Tennessee. Two
decades later similar islands of unpeopled areas were
found in the Ozark Mountains of northern Arkansas

and southern Missouri, where a rugged hill country with poor soil repelled settlement. Vacant spots of this class began to contract and eventually were filled up, but their scars are still left in the sparsity and retarded development of their populations even to-day. Other vacant spots also were due to geographical causes, — swamps, as in northwestern Ohio, western Indiana, southern Georgia, and along the Gulf coast of the southern states; flood-plains in the lower course of several Mississippi tributaries, like the St. Francis, White, and Yazoo, which periodic overflows and continual malaria made unfit for settlement until an accumulation of capital and men justified the construction of levees. Here, too, the scars have lingered in the form of a sparser population. Elsewhere areas of severe climate, small streams unfit for steam navigation, and dense forests rendering the preparation of the ground for agriculture more difficult, were long left uninhabited. These causes checked the landward advance of population in Maine, especially after the decline of wooden ships; and for the same reasons, further west, the incoming tide of settlement set towards Michigan and Wisconsin only to forty-three degrees north latitude, and then was deflected towards the trans-Mississippi country with its fertile prairies and navigable streams. Much of this northern forest region is empty to-day except for an occasional winter logging-camp or the hut of the summer sportsman.

The other vacant spots on the maps showing density of population from 1820 to 1840 were due to the presence of Indian tribes who held large tracts of fertile land right in the natural path of expansion of some of the oldest settlements. Choctaws and Chickasaws prevented Tennessee settlement from extending to the Mississippi River, and kept the population of Mississippi crowded down in the southwest corner of the territory. Further east the Cherokees and Creeks blocked Georgia's advance and crowded upon the flank of Alabama; though the latter state, placed between these two great Indian areas, had developed more rapidly during its short existence than its two neighbors, Georgia and Mississippi, which enjoyed far greater geographical advantages. Georgia's population, dammed up against this Indian barrier, overflowed into the valley of the Chattahoochee and moved down its channel into Florida. In northern Illinois, the Sac, Fox, and Pottawatomie tribes delayed settlement in large areas of fertile land which they held.

The decade from 1830 to 1840 saw the title to all these Indian lands gradually extinguished by the federal government and the tribes themselves removed to the Indian Territory. Within two or three years thereafter, the area thus relinquished was covered with a comparatively dense population, and in the next decade was as thickly settled as any parts of these states. Here there were no scars because no adverse natural conditions.[2]

The elements of population moving into the Northwest Territory, as likewise their distribution, were determined largely by its two great geographical lines of approach. New England, New York, New Jersey, and Pennsylvania furnished most of the early and later settlers; these entered the country by the Ohio River and

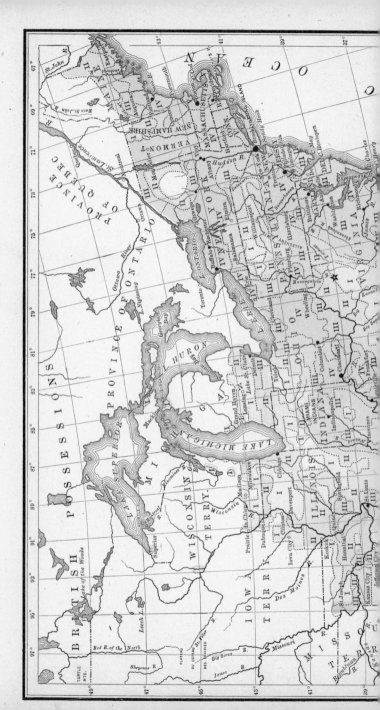

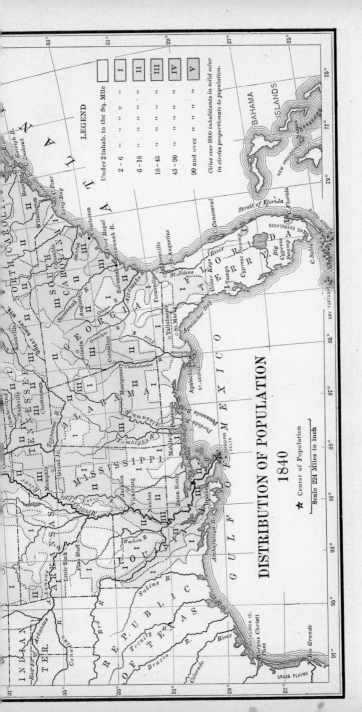

DISTRIBUTION OF POPULATION
1840

★ Center of Population

Scale 224 Miles to inch

LEGEND

Under 2 inhab. to the Sq. Mile	
2 – 6 " " " "	I
6 – 18 " " " "	II
18 – 45 " " " "	III
45 – 90 " " " "	IV
90 and over " " " "	V

Cities over 8000 inhabitants in solid color in circles proportionate to population.

spread northward along its tributaries towards the watershed, or followed the route of the Mohawk depression into the northern part of Ohio, stretching a band of settlement along the whole southern shore of Lake Erie as early as 1820. But beyond, the northern parts of Ohio, Indiana, and Illinois were cut off from the influx of population coming from Lake Erie because of the swamps covering the uncertain watershed of the Maumee, Wabash, and St. Joseph rivers. The southern parts of these states received considerable accessions of population from neighboring Kentucky and Tennessee. Some of the southern immigrants with anti-slavery sympathies were attracted thither by the free-state clause in the Ordinance of 1787; others by the chronic land-hunger of the day.

The geographical position of Illinois was interesting in 1818. With the exception of a group of fur-traders at the lonely station on the Chicago River, all its population was concentrated in the extreme southern part, where focused the Missouri, Mississippi, the Ohio, Cumberland, and Tennessee rivers, and with them the influence of the South and West. Here was a strategic area for military and political purposes. In 1818 Illinois applied for entrance as a state. The Ordinance of 1787 provided that the northern boundary of the north Ohio states was to follow a line drawn through the southern bend of Lake Michigan, thus excluding from Illinois the post at Chicago which was to be the outlet of the already projected Illinois and Michigan Canal. The withdrawal of the western and southern states from the Union was even then a possibility ever present to men's minds.

Therefore it was decided to give Illinois a goodly shore line on Lake Michigan in order to strengthen its northern line of connection, which the Erie Canal was relied on to perfect, and thereby weaken its dependence upon the South and West. Hence a politico-geographical necessity traced the northern boundary of Illinois.[3]

The opening of the Erie Canal in 1825 exerted a strong influence throughout the Lake region, and soon afterwards other canals across the low watersheds southward to the Ohio opened up a hitherto inaccessible country. Into this region, especially after the removal of the Indians to the west, poured a stream of immigrants from the eastern states and from Europe, especially Germany. Between 1820 and 1840 the states north of the Ohio and east of the Mississippi increased as a whole over 360 per cent.[4] The frontier of continuous settlement outlined an almost wholly compact area along the forty-third degree of latitude; while isolated cabins and farms beaded the shore of Lake Michigan and indicated the future line of expansion.

The New York, New England, and German elements brought with them the staid, contracted ideals of the old Atlantic seaboard and of Europe; they tended to settle on moderate and equal-sized farms on the uplands between the streams, put solid improvements on their land, reflecting their sedentary purpose, while only a slow, compact protrusion of their frontier registered their advance at the cost of the wilderness. The trans-Mississippi Westerners, on the other hand, with the half nomadic instinct bred from many migrations, and inherited from their Virginia and Carolina forbears, strung out their settlements along watercourses whence a fresh movement was easy, put up temporary buildings which could be abandoned without a pang,[5] left the upland region between unoccupied, and strode forward into the up-river country with the glory of the sunset in their faces and the yearning for the bigness of the prairies in their hearts.

Expansion between the Ohio and the Gulf went on rather more rapidly than in the Northwest Territory. The small group of settlements between the Natchez District and the Yazoo delta formed the nucleus of the state of Mississippi, as those on the Tombigbee did of Alabama. The purchase of Louisiana attracted immigrants to this region, but they were forced to take the long roundabout river journey by the Ohio or Tennessee and Mississippi. And such as came this way were generally from North Carolina, Virginia, Kentucky, and Tennessee. Accessions from the nearer source of Georgia and South Carolina were excluded by the barrier of the hostile Creeks, until Jackson's victory in 1814 rendered travel across their country less dangerous, and the natural highway around the southern end of the Appalachians could come into use. Hence by 1820 the Tombigbee settlements had grown and spread till they coalesced with those on the bend of the Tennessee and the expanding area of civilization in southern Mississippi.

Kentucky and Tennessee at this time were cut off from the Mississippi by a wedge of the Chickasaw tribe which extended as far as the Ohio between the Mississippi and the Tennessee. But in 1820 the claim of the

Indians was extinguished and immigrants poured into the Chickasaw Purchase, and soon by the Obion, Forked Deer, Hatchie, and Wolf rivers reached the long desired lands on the Mississippi. The southern boundary of Tennessee (35° N. L.) marked the limit of the Chickasaw Purchase, so expansion was checked at this line till 1838, when all the southern tribes were removed to the Indian Territory. Then from the already densely populated Chickasaw Purchase, immigrants rushed into northern Mississippi. The new lands thus opened were peopled chiefly by their neighbors, Georgia, the Carolinas, and more particularly Tennessee, which has been called the "mother of states," because she contributed more to the new settlements in the Mississippi valley than any other state in the Union. From her came an important part of the population of northern Alabama, Mississippi, and Florida, as also a large portion of the early settlers of Missouri, Arkansas, and Texas.[6] To all of these states she holds a central position, and from her borders radiate the rivers making them accessible.

It was in the trans-Mississippi country, however, and especially in Missouri, that expansion was most active from 1810 to 1840. To the unbounded West came all those restless spirits in whom opportunities for migration had aroused all the *Wanderlust* long dormant from the opiate of a sedentary civilization, and to whom movement had become a habit of life. On they came, the more vigorous and daring pushing on to the western rim of the Mississippi valley to set their beaver-traps on the head streams of the Pacific rivers; the less daring taking up "a bunch" of prairie land beyond the outskirts of settlement on which to graze their cattle, and erecting a temporary shelter to be occupied till a more industrious and permanent race of agricultural settlers began to press upon their rear. Then they moved on, — hunting, trapping, rounding up their cattle, planting an occasional field, but always more assimilated to the wilderness beyond them than to the civilization behind. The greater part of the population belonged to this class. The West was peopled by a sort of staccato movement. Immigrants built cabins and laid fences; but before the wood was seasoned or the fresh rails darkened by exposure, the place was sold,[7] and the wheel-tracks of the canvas-covered wagon going to meet the advancing afternoon shadows told the wide destination of the migrant family. Thus the frontier zone broadened until it reached the margin of the arid belt; then a partial recoil set in.

Along the Mississippi boundary of Missouri was the gathering of the waters; hither turned the Ohio with its vast tributaries, the Illinois, and the Missouri. The Mississippi furnished communication with the Gulf below, and with the Green Bay and Chicago portages above. Missouri was accessible to American settlers from the Lakes to the Gulf, and straight across its broad territory ran the natural highway of its own mighty stream. Every section of the United States, therefore, contributed to the population of this nascent state, though some sections more than others.

The old trading intercourse between the early trans-Allegheny commonwealths and the French settlements on the west bank of the Mississippi made the Missouri

country the natural destination of restless backwoodsmen like Daniel Boone and his sons. The right of
property in slaves, guaranteed by the Purchase treaty to
the original French inhabitants of Louisiana, fixed Missouri's future as a slave state, and enabled it to receive
accessions of population from all the other slave states
which the Ohio could deposit at its doors. Its chief
sources of population were therefore Kentucky, Tennessee, Virginia, and North Carolina ; the states just
north of the Ohio, also, yielded emigrants who were
lured westward by the current of " La belle Rivière." [8]
Other settlers came from New England by way of the
Erie Canal, the Great Lakes, and the Illinois River ; and
in 1833 came some thirty thousand frugal and industrious Germans.[9]   As early as 1820 Flint had found in
the Girardeau district a substantial settlement of Germans, some of whom had come directly from the Father-
land, but the greater portion from Pennsylvania and
North Carolina.[10]   After the introduction of steam navigation, foreign emigrants began to land at New Orleans
and ascend the Mississippi by steamboat to this western
country,[11] while they poured in by the thousands from
the northern waterway of the Lakes.

In the distribution of population in Missouri, the geographic control of the rivers was the dominant factor
up to 1830. Settlements spread in a narrow belt up the
Mississippi and the Des Moines to the northern boundary
of the state ; in a broader zone up the Missouri, chiefly
on its northern or prairie side, to the western frontier
defined by the elbow of this great river ; and in a still
broader band along the small western affluents of the

Mississippi, especially the Meramec, and the head streams of the Whitewater and St. Francis, just above the swamps in which these rivers lose themselves in the southeastern part of the state. The hilly country of the upper Whitewater and St. Francis extends eastward as the Cape Girardeau country, early selected by the French for settlement; it is fertile, thickly timbered, well watered, furnishing pure streams and mill-sites, and for these reasons early attracted colonists while the still richer alluvial bottoms remained untouched.[12] Mill-sites were a rarity in this level country. For a long time maize was pounded in a mortar by hand. In 1825 steam-mills were introduced in St. Louis, but elsewhere treadmills worked by oxen or horses did the work,[13] and lumber was costly because of the lack of water-power sawmills.

At a time when only the rivers with their rich bottoms, flanking prairies, and natural highways, could boast a population of from two to six to the square mile, the ranger and the rancher, nevertheless, were setting up cabins and fences in remoter parts of Missouri. Flint says that as early as 1818 there were white settlers in the wilderness near the western state line between the Arkansas River and Missouri, a hundred miles from any settlement. Thither they had gone to secure a fresh range for their cattle, or merely the " elbow-room " which was the first demand of the backwoodsman.

By 1840 population was distributed evenly but sparsely over the whole state except in the rough hill country of the south-central portion. A uniformly fertile soil resting on an underlying limestone rock, abundant rainfall, temperate climate, a country too level to present any obstacle to expansion, either on its surface or in the current of its streams, diversified by forest and prairie, the natural focus of the Illinois, Wabash, Ohio, Tennessee, and Mississippi river routes,— no wonder that Missouri increased in population. Here grew the whole range of our northern agricultural crops, and the labor, so difficult to secure in a new country, was here supplied in part by slaves. The staple crops were wheat and hemp; the former was sent down the Mississippi in large quantities as flour to New Orleans, and the latter, either as raw material or manufactured into bagging and rope, was exported to Kentucky.[14]   St. Louis became the great emporium for the commerce and manufacture of the upper Mississippi, the depot of the western fur trade which the Missouri River system brought to its doors, and the eastern terminus of the Santa Fé trade which was carried on via the Missouri River to Independence, and thence by means of caravan across the American desert. St. Louis, by its command of the great inland water routes, early developed the manufacture of steamboat supplies, steam machinery, and finally by 1844 rivaled Pittsburg, Cincinnati, and Louisville in the construction of superior steamboats.[15]

As Missouri filled up, and as settlements between 1830 and 1840 were spreading from northern Illinois and from the southwestern nucleus of that state, the natural expansion of these three areas of population was creating the beginning of Iowa. No artificial state line or the demarking course of the Mississippi ever really separated the settlements distinguished for administrative purposes as Illinois, Missouri, and Iowa Territory. In 1832 the government purchased from the Indians nearly all the land within fifty miles of the west bank of the Mississippi, from the Des Moines to the mouth of the Wisconsin River on the north. The beautiful, fertile prairie country was immediately taken up by pioneers. Even in 1830 Illinois and Missouri had overflowed into the little angle between the Des Moines and Mississippi rivers; but now settlement spread rapidly north to Burlington and Salem, and by 1840 this southeast corner of Iowa was the most densely populated portion of the territory. It merged into a belt of similar density, which extended up the Illinois River and over to the end of Lake Michigan, thereby suggesting the avenue of approach.

At about the same time as the planting of Burlington (1833), Dubuque was founded nearly two hundred miles farther up the Mississippi, and became a center of distribution for the northern part of the newly purchased area. Its location coincided with the western end of a band of denser settlement which defined a broad belt of rich lead-mines extending along the northern border of Illinois and over into southwestern Wisconsin. These mines had been worked by Indians and French traders from early times, but the rush thither began in 1823, when they were described as the richest mines in the world. The chief diggings were at Galena and the vicinity in Illinois, and were pretty well distributed over the present counties of Grant,

Lafayette, and Iowa in Wisconsin, while rich mines were found across the Mississippi about Dubuque.[16] Hence the birth of this city followed closely on that of Galena in 1826.

Population advanced rapidly up the drainage streams of Iowa, covered the whole area of the Black Hawk Purchase by 1836, and was encroaching on the Indians' lands to the west, so that the government in 1837 extended the western boundary by a new purchase from the Sacs and Foxes. A new and strong tide of immigration, begun in 1839, carried population far up the Red Cedar and Des Moines rivers, and on the western frontier up the Missouri beyond the present site of Council Bluffs, while a vacant spot between the eastern and western groups showed the geographic control of the Iowa rivers in determining the location of settlements.

As Iowa lay north of the Missouri Compromise line, the movement thither was from free states. The route of the Erie Canal and the Great Lakes brought immigrants from New York and New England, and foreign elements from Germany, France, and Great Britain,[17] who, with their small capital, could find no place in the industrial system of the South. The Ohio also delivered its contribution of settlers from the free states on its northern bank.

The movement of population, at least after 1820, was in general from free state to free, and from slave state to slave. This was true in the Northwest Territory and in Iowa, as we have seen. Missouri, occupying an intermediate position between the two sections and convenient to the Ohio and the northern routes, drew from both sections, but chiefly from the region south of the Ohio. Arkansas and Louisiana by their geographical location belonged to the slave power. They recruited their population from the older and more densely settled commonwealths of Kentucky and Tennessee, and from the wealthy planters of Mississippi, Alabama, South Carolina, and Georgia, who were seeking new, unexhausted lands for the employment of their slaves. These, therefore, generally crowded into the rich bottoms where cotton, sugar, and maize would flourish.

Emigration to Arkansas went on very slowly till after the Indian Territory was erected as a separate district (1824), but from 1830 till 1835 the population doubled itself. This was the period when Mississippi, Alabama, and Georgia were feeling their expansion blocked by the presence of the Indians in the extensive areas within their boundaries, and when the final purchase and removal of 1838 had not yet brought them relief. Louisiana, being an older state, had less unclaimed land to offer the newcomer than her neighbor to the north.

The population of Arkansas in 1820 was 10,000. These were distributed in one narrow, detached line of settlement along the Mississippi near the mouth of the St. Francis, and in another fifty miles below the mouth of the Arkansas; in a broader line following the Arkansas upward from a point fifty-five miles above its confluence with the Mississippi, the area of cypress-swamp, bayou-network, and ox-bow lakes which told the story of widespread and long-continued overflow.[18]   This first point of safe habitation — leaving out of consideration ague and fever — was occupied by the Post of Arkansas, in early days the administrative center of the territory. A line of continuous settlement extended from here to about forty miles above the present site of Little Rock ; but beyond that, far up the Arkansas, the " Mulberry Settlement " in the hill country of the western frontier told of the energy of the pioneer and his desire to get out of the malarial bottoms.   Two other detached settlements indicated the same need, one in the north among the highlands of the White River, and another in the southwest, known as Mount Prairie, on the tableland between the main Washita and its tributary, the little Missouri.[19]

With the new accessions of population, these centers expanded till at certain points they coalesced, but elsewhere were separated by vacant areas of mountain, swamp, and flood land, while the line of the Arkansas valley rapidly carried settlement to the pleasant hill country in the northwest, bordering on the Indian Territory.   In 1835 rich planters were attracted to the southwest corner of the territory where the deep meanders of the Red River furnished rich bottoms suited to cotton and maize.[20]

The soil along all the great streams of Arkansas was the richest alluvium and practically inexhaustible.   The extensive prairies and hill country also, by the decay of their bedrocks, had an excellent soil, which eventually was found to yield good crops of cotton, but in the pioneer days of the state was taken up by the grazing element. A semi-tropical climate, abundant rainfall, and accessibility to the Mississippi secured to Arkansas both crops and a market.

Similar geographical conditions and similar results obtained in Louisiana. Here too were swamp, alluvium, prairie, and pine woods. " The people in the pine woods raise cattle by the hundreds and thousands, — are poor, satisfied, and healthy. In the bottoms are the sugar and cotton plantations with wealth and sickness." The prairie country west of Opelousas and the upper Bayou Teche was devoted chiefly to cattle, partly to cotton and sugar plantations. The lowlands were mainly swamps, so that we find the valley population distributed in belts from two to three miles wide along the elevated margins or broad nature-made dikes of the watercourses. Here, as in the river peninsulas of old Virginia, the planters were able to load their produce of sugar, molasses, and cotton immediately on steamers at their own wharves. This is Louisiana as Flint saw it in 1823.[21] In 1836 a fresh tide of settlers, avoiding the swamp regions of the coast, spread into the arable lands southwest of the Bayou Teche and along the head waters of the small Gulf streams as far as the Sabine. This represented chiefly a natural expansion southward of the Red River zone of settlement, and it was supplemented by an advance of planters at this time into the fine cotton region of the upper Red, Washita, Black, and Tensas rivers, all in the northern part of the state.[22] The settled areas along the Mississippi and the Red now merged into those of Arkansas, as shown in the census map of 1840. The abrupt line defining the northwestern frontier of Louisiana settlement represents the political separateness of the country beyond, but it conceals the important fact that the expansion of the American people, though not of the American government, had gone far beyond the limits of the Sabine.

The subject of early trans-Mississippi expansion must include that irregular advance of American citizens into the domain of Mexico. We have seen that expansion into Spanish Louisiana foreboded the political fate of that vast territory, which, however, by the rapid shiftings of diplomacy and the clamorous demand of the youthful West, reached its destiny without waiting for the slow operation of the natural law of territorial extension. Thirty years later in Mexico this law, reinforced by other politico-geographical principles, worked itself out to its inevitable conclusion, just as it has done with slight local variations more recently in the Hawaiian Islands.

The advance into the Texas province of Mexico was a part of the general westward movement. It was stimulated by the restlessness, enterprise, and aggressive spirit of the frontier; and it was participated in largely by the slaveholders of the southwestern states whose legitimate political expansion was blocked by the Indian Territory and the Missouri Compromise line, but it also drew into its sweeping tide representatives from every state in the Union and emigrants from across the sea. To the landless, the wide grassy plains and fertile bottoms of Texas were attraction enough. To the merchant or trader, Mexico with its belated industries and rudimentary commerce would yield a fortune in a few expeditions.   The uncertain claim of the United States government to the trans-Sabine country before the final understanding of 1819, the well-known weakness of the far-away power in Madrid, the attraction of an ill-controlled border region of a foreign country to the rough, lawless elements of our own frontier, probably all united to swell the American population in Texas.

As we have shown before, a frontier is always a zone of assimilation between the contiguous territories.   It will lean more to the one or the other side of the political boundary line, according to the energy and growth of the national life behind it.   We have seen how, just after the Louisiana Purchase, Spain, by the energy she displayed in strengthening her hold on the Texas country by settlement, fort, and garrison, secured the Sabine instead of the Rio Grande as the eastern frontier of Mexico.   The United States had not at that time accumulated a large enough population in the southwest to constitute a natural pressure upon the Mexican frontier.   From that time, Spain, seeing the unwisdom of her previous policy of hospitality to American settlers, strove to keep the Texas district uninhabited on the primitive principle of a waste boundary as a barrier against the influx of her restless neighbors.   But the principle did not work.   It is easier to occupy a vacant land than an inhabited district.   On came the Americans, and the zone of assimilation grew wider and wider, extending farther and farther on the Mexican side of the Sabine.

The aggression of the Americans was shown by their share in almost every revolution in Mexico between

1812 and 1817. Officers of the United States army helped to lead them, and the expeditions supporting them were fitted out at Natchez on the Mississippi, Natchitoches on the upper Red River, and Gaines Ferry on the Sabine. When the treaty of 1819 renounced the claim of the United States to the territory between the Sabine and the Rio Grande, American interests in Texas were already so strong that Henry Clay protested against the agreement. Natchez, from whose doors the Red River led to the Texas frontier, entered her protest against the treaty by sending out an expedition under Dr. James Long to invade Texas. His small force of seventy-five men swelled to three hundred by the time he reached Nacogdoches, fifty miles west of the Sabine. There he issued a Declaration of Independence for Texas, but his enterprise failed of its purpose.[23] In 1820 all Mexico revolted against Spain, and in four years the United States of Mexico was established. At this time it included in its population about three thousand Americans, settled mostly at Nacogdoches, the first town on the highway leading from Natchitoches, and along the road southwest to San Antonio.

Mexico, anxious to increase her population, now met the spirit of expansion in the American people by issuing large grants of land in Texas. Stephen Austin, originally from Connecticut but more recently from Missouri, secured an enormous tract extending from the highroad between Nacogdoches and San Antonio to the coast between Galveston and Matagorda bays, on which, as *empresario*, he was to settle three hundred families.[24]

Other contractors from the trans-Allegheny states, Missouri, Tennessee, Kentucky, and Ohio, or from New York, Ireland, and Scotland, agreed to bring some two thousand families, and received similar grants until almost the entire state was parceled out. This was in 1825; by 1830 there were twenty thousand Americans in Texas. They came by all the ports of the coast and moved up the rivers; or they followed the Ohio and the Mississippi to the Red, which carried them up to the head of steamboat navigation at Natchitoches, or further still by smaller craft to Shreveport, somewhat nearer the Texas frontier. The eastern border of Texas about the Iyish Bayou was chiefly settled by Americans,[25] and beyond lay the rich boundless prairie on which could be raised fine crops of sugar, cotton, and corn. The genial, salubrious climate, too, had a peculiar attraction for the ague-shaken dwellers of the Mississippi lowlands.

Four roads, indicating the amount of intercourse between Texas and the United States, crossed the frontier in 1826, — one from Liberty on the Trinidad to Opelousas, Louisiana; another from Liberty, joined at the Sabine by a branch road from Nacogdoches, leading across to Alexandria on the Red River; the old highway direct from Nacogdoches to Natchitoches; and a fourth from the former town across the Red River at Fulton to Little Rock, Arkansas.[26]

Mexico became alarmed and withdrew her favors, canceled all except three land grants, forbade further colonization and the importation of slaves; and to check the active trade between the two countries, closed all

but one port on the American side and laid heavy
import duties on all the manufactured articles needed
by the agricultural population of Texas. In the mean
time the spirit of expansion was making itself manifest
in another way. The United States government was
becoming desirous of acquiring this last territory occupied by the American people, as it had followed them into
Louisiana and into West Florida. From 1827 to 1829
repeated offers were made by the government at Washington to buy Texas. The reasons advanced were
politico-geographical ones. Mexico was too near to
New Orleans, the entrepôt of the great American waterway. Much of the Red River and many tributaries of
the Arkansas lay in Mexican territory. When the
country should become thickly settled, community of
interest in the navigation of these streams would give
rise to disputes. Clay suggested the Brazos, or the
Colorado, or the Rio Grande with the western watershed of the Red and Arkansas rivers as boundary; and
when these offers were rejected, he warned Mexico of
the possibility of her losing Texas because of the large
American contingent there, and the consequent collisions
between the two elements of the population which might
draw the two republics into war.[27]

Finally the Texas uprising came and a new star
dawned for the American flag. The revolt of Mexico
against Spain, like the American Revolution, was the
repudiation of the central authority by a peripheral state
according to the politico-geographical law already stated.
Somewhat similar, though somewhat different, too, was
the declaration of independence by Texas. In the

American colonies there had been little infusion of foreign blood to modify the race : a pure-blood offspring rebelled against parental authority. In Mexico, the ethnic element of the Spanish conquerors had been greatly weakened in its absorption into the large native population. Hence racial differences had combined with new and unfamiliar geographical conditions to differentiate the Mexican from the Spanish, and strengthen the tendency towards disintegration. Texas, too, was a peripheral state, with all its natural tendencies towards defection enhanced by the fact that the dominant race here — dominant by reason of energy, intelligence, wealth, and affiliations, though not by number — was of an imported, alien stock, having no inherited sympathy with the governing power.

The geographical location of Texas on the outskirts of the Mexican territory, remote from the center of the federal authority, with a savage frontier on the north, and an alert, enticing neighbor on the east, put this province much in the situation of the early trans-Allegheny settlements in relation to their mother states of Virginia and North Carolina, and generated the same separatist tendencies with the same threat of defection to a neighboring foreign power. It was difficult to get a hearing before the general government. " Texas is situated twelve hundred miles from the capital of Mexico ; and owing to the distance, and state of war in the country ever since we have known it as a Republic, communication could be no other than tedious and uncertain. . . . All these . . . have been sufficient to cause so much delay as to injure Texas materially without the possibility of remedy." This is an extract from a circular issued in Texas in 1834, advocating patience and loyalty to the Mexican government. Two years before this Texas had sent in a petition for independent statehood and separation from the more densely populated province of Coahuila extending south of the Rio Grande, on the ground that Indian problems along her northern frontier threatened a calamity "which nothing short of a well-regulated government of a free, unshackled, and independent state can provide against." "The wide extent of wilderness, forming a natural boundary between Texas and Coahuila, places an indispensable barrier in the way of Coahuila's extending the efficient means of defense she might wish." [28] Furthermore, differences of climate, soil, and productions, partly also of population, made such a separation desirable. This line of reasoning might have emanated from any of the pioneer leaders of the Watauga, Cumberland, or Kentucky settlements.

Remoteness in turn enabled the Texans to ignore certain Mexican laws. A long reach weakened the arm of the executive here as we shall find was also the case in Mexican California. The inhibition of the slave-trade was disregarded by the Americans in Texas, as was later (1829) the decree of emancipation of all slaves; and the danger of revolt attending the enforcement of this law was so apparent that a special decree made it inapplicative in this province. Even in the days of Spanish supremacy in Mexico, the connection between Texas and the center had been weak. Control of the province had been administered after a fashion from the remote

Chihuahua, five hundred miles even from San Antonio in whose vicinity was the only compact area of Spanish settlement west of the Sabine and Red rivers. Subsequently the vacant region between these border streams and the Rio Grande was filled by a foreign population, prone to fall apart from the Mexican center by the weight of their own antagonisms, and furthermore drawn in the opposite direction towards the United States by every tie of race and interest. Here attraction on the one side and repulsion on the other operated to the same end.

When the revolt of Texas began, its affiliations with the United States counted for something. Two companies of New Orleans volunteers, others from Mobile and Kentucky, — some eight hundred in all, — took a hand in establishing the young republic, while two hundred deserters from the United States army were found serving under the Texas standards; and Sam Houston and Davy Crockett, men who embodied the enterprise and recklessness of the backwoods frontier, made picturesque figures in the Texas war of independence.

How truly this was in the last analysis a war of unauthorized or individual American expansion, is indicated by the fact that, of the fifty-eight delegates who declared the independence of Texas at New Washington in March, 1836, only three were Mexicans, though Americans constituted only about one fourth of the total population. Their immediate application for annexation to the United States showed that in their own consciousness they had never really expatriated themselves, but merely had stretched the farming, cattle-tending, trading America out over Texas prairies, with the expectation that the flag would do its smaller part and follow. Therefore, when the admission of the Lone Star State was agitated ten years later, the American mind reverted to the old claim of the United States to the Rio Grande boundary of Louisiana, and argument ran high for the "reannexation of Texas."[29]

NOTES TO CHAPTER IX

1. Eleventh Census of the United States Population, vol. i. p. xxii.
2. For general subject of vacant areas, see ibid. p. xxviii. Also, Treaties of the United States with the Several Indian Tribes, 1778–1837, pp. 391, 497, 633. Washington, 1837.
3. Ford, History of Illinois, pp. 22–24. Chicago, 1854.
4. Sparks, Expansion of the American People, p. 268.
5. Flint, The Last Ten Years in the Valley of the Mississippi, p. 48. Boston, 1826.
6. Monette, History of the Valley of the Mississippi, vol. ii. p. 286. 1846.
7. Flint, The Last Ten Years in the Valley of the Mississippi, p. 206. 1826.
8. Monette, History of the Valley of the Mississippi, vol. ii. p. 553.
9. Ibid. p. 554.
10. Flint, The Last Ten Years in the Valley of the Mississippi, p. 233.
11. Brownell, History of Immigration ; see New Orleans in list of ports between 1820 and 1840.
    Monette, History of the Valley of the Mississippi, vol. ii. p. 567.
12. Flint, The Last Ten Years in the Valley of the Mississippi, p. 232.
13. Ibid. p. 211.
14. Monette, History of the Valley of the Mississippi, vol. ii. pp. 553, 554.
15. Ibid. p. 558.
16. J. M. Peck, A Guide for Emigrants in Illinois and Missouri, pp. 132–134. Boston, 1831.
17. Monette, History of the Valley of the Mississippi, vol. ii. p. 568.
18. Flint, The Last Ten Years in the Valley of the Mississippi, pp. 252–254. 1826.
19. Ibid. pp. 265, 266.

20. Monette, History of the Valley of the Mississippi, vol. ii. p. 555.

21. Flint, The Last Ten Years in the Valley of the Mississippi, pp. 329–349. 1826.

22. Monette, History of the Valley of the Mississippi, vol. ii. pp. 517, 518.

23. McMaster, History of the People of the United States, vol. v. pp. 4–7. 1900.

24. Ibid. p. 13.

25. Flint, History and Geography of the Mississippi Valley, vol. i. p. 464. 1832.

26. McMaster, History of the People of the United States, vol. v. p. 12. Map of the period, 1900.

27. Ibid. p. 461.

28. David B. Edward, History of Texas ; or the Emigrant's, Farmer's, and Politician's Guide, pp. 231 and 207. Cincinnati, 1836.

29. Sparks, Expansion of the American People, p. 318.

CHAPTER X

GEOGRAPHIC CONTROL OF EXPANSION INTO THE FAR WEST : THE SOUTHERN ROUTES

The westward moving frontier of the American people is beyond all doubt the most interesting subject that American history presents. Here is written the fullness of American energy, its daring resourcefulness and ambition ; here the rate of national growth registers itself in more telling figures than mere statistics of population ; here, with rifle, axe, and plough ; with canoe, bullskin boat, and pack-horse, the man of backwoods and plain shapes the national dream of empire into the sturdy stuff of trading-post and ranch and farm.

A frontier is a zone, and the width and character of that zone tells the whole story. A narrow, evenly drawn frontier, like that of the American colonies in 1750 along the eastern foot of the Appalachian barrier, points to the balance maintained between the half-developed strength of the nascent people and the power of British dominion reinforced by the geographical control of the mountain wall. In 1800 the frontier has lost its neat outline but has gained in interest. It sends out great bulges and streamers running out from the Appalachian ridges to Lake Ontario and to Erie, and to the Wabash. Its width bespeaks a rapid rate of expansion, while farther still, outlying groups of settlement in the Natchez District or opposite the Missouri's mouth, and lonely cabins beyond the Mississippi in Spanish territory lead the advance and signal to the rear to follow over the easy roads which river and prairie afford.

In 1820 and 1830 the western frontier is more ragged still and forms a fringe with wide intervals up the trans-Mississippi rivers. The map of 1840 shows what seems to be again the neat, narrow frontier zone of arrested growth, a line approaching the ninety-fifth meridian and the northward bend of the Missouri River. Though continuous settlement did pause long at this limit, because it was the outer margin of the arid belt and the eastern boundary 'of the Indian Territory, nevertheless even before this time, that unofficial America on Texas soil was supreme as far as the Nueces River. American trading-posts at the gates of the Rockies, missions on the Columbia, and ranches on the Willamette had stretched the United States frontier to the Pacific, while individual enterprise was Americanizing the commerce of New Mexico and influencing the politics of Alta California. Never were the accumulated energies of the American people so great, never was its frontier so broad a zone. Over arid plain, snow-capped mountain, and alkaline desert, it stretched from the Missouri to the Pacific, wherever the smoke from the trapper's camp-fire in the Rockies curled upward in the evening air, wherever the trader hobbled his footsore beasts at night in the scant meadows of the western trails, wherever the immigrant staked out his land claim on the banks of the Columbia or the Sacramento. The breadth of this frontier explains the fact that, while continuous settlement paused at the western boundary of Missouri and Iowa in 1850, the United States had made good its claim as far as Puget Sound, and the American flag waved over the presidios of California.

To the first stage of this far western expansion nature presented no serious obstacles, though she also afforded no great assistance, such as that rendered by the Ohio River and the Great Lakes in the trans-Allegheny advance. The even, gradual rise of the Great Plains to the foot of the Rocky Mountains made the ascent to an altitude of even five thousand feet almost imperceptible. No forests obstructed progress, and the character of the soil was such that the passage of a few wagons marked out a well-beaten road. The rivers draining this even slope flowed in parallel courses to the Missouri and Mississippi. They pointed the direct way to the West, and were closely followed by the moving caravans for the sake of their water supply. The cottonwoods growing on their islands or scantily fringing their banks furnished fuel for camp-fires, which otherwise had to be supplied by *bois de vache*, the " buffalo chips " of the plains. The arid and treeless regions of the world everywhere resort to similar fuel, — the excrement of camels in the deserts of Arabia, and on the rainless plateau of Tibet, cattle dung, with which Kipling's hero, the boy Kim, also cooks his evening meal in the deforested valley of the Indus.

With the exception of the Missouri, the western rivers afforded no waterways to the emigrant, for they are navigable only in rare, short periods, and only for

200 MILES TO INCH
TRANS-ROCKY TRAILS ....
LEWIS AND CLARK ROUTE — — —
LEWIS RETURN ROUTE >>>>
CLARK RETURN ROUTE ++++
PASSES ✕✕✕

Copyright, 1895, by Alex. E. Frye.
From Frye's Geography, by permission of Ginn & Co. Publishers.

MAP OF

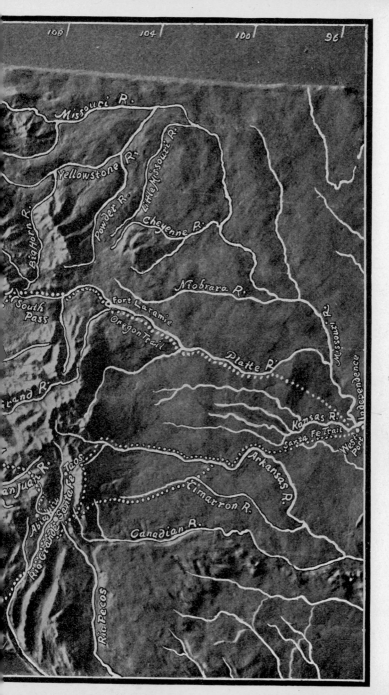

canoes or very shallow flatboats. Rising in the Rockies they flow directly east through an arid land, receiving no tributaries to increase their volume, while their water is evaporated in the dry air or sucked up by the porous soil. In crossing the broad belts between these rivers the early caravans would often travel a hundred miles without meeting a single permanent watercourse. Hence the trails preferably passed from one river to another somewhere between the ninety-fifth and ninety-eighth meridian on the eastern margin of the arid region, and again by the lateral streams at the base of the Rockies where the beds were still full.

These rivers, owing to their usually scanty volume and consequently weakened currents, possess only slight erosive power; hence they have deposited much detritus in the valleys previously cut in the friable soil of the prairie. Over these sandy plains in time of flood they spread a shallow sheet of water, often two miles wide; but when the flood passes, the river becomes a series of shallow interbraided streams, obstructed by sand-bars and islands. The topography of the country gives a different character to the Missouri. From 48° to 39° north latitude this stream is a lesser Mississippi, occupying a central basin running north and south, and intercepting all the western streams which set out from the mountains on their way to join the Father of Waters. Hence the Missouri is reinforced constantly by new affluents. Moreover, it rises in the extreme north of the Cordilleran plateau, where the mountain system is narrower and in general lower than it is further south, thus admitting the moisture drifted in by the Pacific westerlies; hence this river has better supplied sources than the Platte and the Arkansas. Nor is this all. The upper Missouri by its tributaries spans the base of the Rockies from the sources of the Milk River (49° N. L.) to the springs of the Big Horn (43° N. L.), while its own main stream penetrates to the western ranges of the system, thus levying a tribute of water on a wide stretch of the highlands. Hence the Missouri is navigable as far as Fort Benton at the mouth of the Teton River, just below where the Great Falls present an obstacle to further advance.

The sources of the Platte, the Arkansas, and the Red stretch out along the eastern slopes of the Rockies with a narrower span, but their lateral head streams interlace with one another, as the North Fork of the Platte does with the southern affluents of the Yellowstone in the Big Horn and Powder River. Thus it is possible to pass from one of these rivers to another by lateral head streams over easy watersheds along the whole eastern base of the Rocky Mountains. These geographical lines determined the trails of the trappers going north and south in the first half of the past century, and later the route of the railroads which now traverse the base of these mountains.

The superior navigability of the Missouri and its course straight from the great fur-fields of the North, naturally drew the canoe of the trader and trapper up its course first. The Lewis and Clark expedition in 1804 had no difficulty in finding Canadian guides and voyageurs who knew the Missouri as far as the Mandan villages at the great northern elbow of the river. On the upstream voyage, the expedition met a trapper, Vallé, who had spent the winter nine hundred miles up the Cheyenne River at the foot of the Black Mountains.[1] It found also French traders among the Mandan villages, and several agents of the Northwest Fur Company who had come from their station on the Assiniboine River only a hundred and fifty miles to the north.[2] The course of the Canadian rivers, the Saskatchewan and Assiniboine, are here parallel with that of the upper Missouri, from which they are separated by only a narrow, lake-dotted watershed; and both systems lead to the springs of the Columbia. These geographical conditions determined the clash of two rival commercial interests on the Pacific coast. No sooner did Lewis find the English trading on the Missouri than he began to speculate whether the Milk, Maria, or some other of its northern affluents might not give the Americans access to Canadian territory on the upper Saskatchewan for trade with the Assiniboine Indians, with whom the English were already doing a lively business.[3]

The advance of the traders up the western streams was rapid. In 1805 the Little Missouri was the remotest point on the Missouri visited by white men; but on his return voyage in 1806, Lewis met two American traders in camp on the White Earth River,[4] and on the lower Missouri he constantly passed strings of canoes on their way to the Platte.[5] Two years later, when Astor's party set out for the mouth of the Columbia to establish their trading-post on the Pacific, Missouri trappers had appropriated all these western streams. They spent the winter among the mountains setting their traps, hunting, and trading with the Indians for buffalo robes ; and in the spring took advantage of the annual rise in the shallow rivers to float their cargoes of furs in canoes or barges down to St. Louis. The flood was generally sufficient to carry them to their destination, but not always. Fremont in 1842 met some traders of the American Fur Company who had started out from Fort Laramie on such a voyage down the North Platte. Their boats drew only nine inches of water, and floated rapidly down the full tide for some sixty miles; but then followed wide, shallow reaches, blind channels, and sand-bars, while the river was falling rapidly. There was nothing to be done but to leave their goods on land with a guard, while the rest, with such packs as they could carry, started out on foot for St. Louis.[6]

When the frontier military posts of the United States were still confined to the western boundary of Missouri and Arkansas, trading-stations, those vedettes of civilization, had been planted far up the head waters of all the western rivers, or beyond the Rockies on the sources of the Pacific streams. As early as 1808 the Missouri Fur Company had a station in the mountains at the three forks of the Missouri, but in 1809 they were dislodged by the hostilities of the Blackfeet Indians, so they crossed the Rockies and established a post on the head waters of the Columbia.[7] The Laramie Fork of the North Platte was marked by two such stations, Fort Platte at its mouth and Fort Laramie[8] a little farther upstream.   Vrain's Fort and Fort Lancaster, two private posts on the South Fork of the Platte, at an

elevation of 5400 feet,[9] gathered in the furs of the
Colorado Rockies, but had rivals just to the south in
Bent's Fort on the upper Arkansas, just above the
mouth of Purgatory River, and Barclay's Fort in the
Mora valley of the Canadian. All of these posts became
later great way stations in the westward movement. In
the wake of the trapper came the pack-horse or wagon-
train of the trader and the white-roofed caravan of the
settler.

In the early days (1810) St. Louis was the last out-
fitting point for the Indian trade of the West; but as
steam navigation was introduced and then improved,
and as the commerce of the prairies increased, the out-
fitting point traveled westward up the Missouri, first
to Franklin, opposite the present town of Boonville, a
hundred miles from the frontier; then, in 1831, to the
new town of Independence, only twelve miles from the
Indian border and two or three miles south of the Missouri; and, finally, when the steamboat landing for
Independence caved in, just beyond to Westport and
Kansas City. These three places then shared the control of the overland trade.[10] The northward bend of
the Missouri here made this elbow the natural terminus of the river route to the West. Here began the
prairie trail. The unfitness of the streams for navigation forced the westward movement to resort to other
means of transportation. Pack-horses and wagon-trains
took the place of canoe and barge and steamboat, dusty
trail over wide grasslands and sandy waste the place of
gliding current twixt wooded banks. Independence
became a market for horses, mules, and oxen. Here the

Rocky Mountain trapper could get his simple equipment; but it was the overland trade with the Mexican town of Santa Fé which gave to Independence its bustling activity and made St. Louis the great commercial city of the early West.

Just beyond the eastern rim of the Rockies in the valley of the upper Rio Grande, and opposite a natural gateway in the mountain barrier, lay the old city of Santa Fé, a territorial capital even in the early Spanish days, and under Mexican rule a squalid adobe town of some three thousand inhabitants. The business of the place was considerable, for it supplied almost the whole population of New Mexico, which was distributed in villages and ranches along the valley of the Rio Grande for a hundred miles to the north of Santa Fé and a hundred and forty to the south.[11] Its markets were stocked under the Spanish régime with goods which came up from Vera Cruz and other Gulf ports of Mexico at enormous cost; but after Mexican independence was established, a less exclusive commercial policy was adopted and the overland trade from the United States began. Then every wagon-track and mule-trail across the plains marked the passing of the shuttle weaving northern Mexico and the American Republic into one fabric. The aridity of the prairies called a temporary halt to agricultural expansion, but wider ranging commerce made Las Vegas and Santa Fé American towns in their business quarters.

A few pioneers in the overland trade between 1805 and 1820 had found their way to Santa Fé by various routes, — up the Platte and the Arkansas to their sources, over the Colorado ranges, and into the valley of the upper Rio Grande above Taos;[12] but the established Santa Fé trade began in 1822, and after a few experiments as to routes, settled down to what was known as the Santa Fé Trail.

Geographical conditions determined Independence, Westport, or Kansas City as the starting-point of the overland commerce. The east and west course of the Missouri between the frontier and the Mississippi made this river the natural western extension of the Ohio route, which brought goods from the manufacturing centers of the country, and for which St. Louis was the natural western depot and distributing-point. North of New Orleans there was no emporium except St. Louis, whose river and land connections with Santa Fé, moreover, were shorter and better than those of the Gulf city. Arkansas, with its exclusively agricultural interests, developed no commercial center which might have profited by the waterway of the lower Arkansas River, tortuous though it was, and the direct western line along the Canadian River to San Miguel and the gates of Santa Fé. This route from the Arkansas frontier was several days shorter than the northern trail, was open earlier in the spring, with a supply of pasturage for horses which lasted, too, later in the fall; it was better provided with timber along the tributary brooks and less intersected by large streams.[13] For these reasons it was sometimes taken by a belated caravan in the winter-time in preference to the colder northern route. But the preponderance of advantages was along the line of the Ohio, the city of St. Louis, the Missouri,

and the upper Arkansas, which therefore determined
the route of the Santa Fé Trail. This road, eight hundred miles in length from the last outfitting-point at
the southern elbow of the Missouri, rising by such imperceptible degrees for three fourths of this distance as
to seem quite level, guided straight across the thirsty
stretches of the Great Plains by the eastward-flowing
streams, bending then slightly southward to the gate in
the mountain wall which opened upon the old Spanish
city, made the simplest and most direct connection between the emporium of the middle Mississippi and the
distributing-center of the upper Rio Grande valley.

The sudden northward bend of the Arkansas between
the ninety-seventh and ninety-eighth meridians brings
the upper course of this river parallel with the Kansas
and almost directly west of Independence, making its
valley the natural supplement by land of the Ohio and
Missouri route to New Mexico. The Santa Fé Trail,
after crossing a well-watered country to the Arkansas,
followed the river westward as far as Bent's Fort
in the early days, turned southwest up Timpas Creek
and the Purgatory River into the Raton Hills, and
over this ridge to the interlocking head streams of the
Canadian and Pecos, which led to Las Vegas and San
Miguel.[14] Fifty miles northwest across the mountains
lay Santa Fé in its hill-locked valley. To the west of
San Miguel the outer range of the Rockies is broken up
into detached plateaus and ridges, between which the
Apache Cañon, a pass three miles long and wide enough
for only one wagon, carried the road over the watershed. Thence with a slight dip it dropped to the

elevated valley of Santa Fé (7000 feet). It was at the
western end of this gap that the Mexican army, in
1846, took its stand to oppose the advance of General
Kearny, who had brought his forces from Fort Leavenworth by this route.

For the first year or two of the Santa Fé trade,
transportation was confined to pack-horses, which could
make their way over the rough hill country at the foot
of the Rockies ; but when the commerce was developed
and wagons were resorted to, the travel left the Arkansas at Cimarron Crossing, just beyond the present Dodge
City, and struck southwest across the desert to the
Cimarron River, which by a decided northern bend here
approaches within fifty miles of the Arkansas. The
road led then southwestward up the half-dry Cimarron,
and struck the older trail at Las Vegas on the Mora,
the most northern affluent of the Canadian River. This
became the established line of the Santa Fé Trail. In
addition to the easier gradients of the plains, it had the
advantage of avoiding the long detour of the upper
Arkansas; but the effort of the pioneer caravan to
cross the desert to the Cimarron proved almost disastrous, and this stretch was always the scene of suffering for man and beast.

Since the Trail ran along the banks of the western
rivers for a great part of the distance, and these, owing
to prevailing aridity, received only a few small tributaries, the drainage system of the country placed no
serious obstacles in this path of overland commerce.
It crossed the unbridged streams by fords in the dry
season, when teams had to be doubled to the wagons

and lashed across the quicksands to prevent their sink-
ing; in spring freshets by buffalo boats and wagons
encased in skins to make them impervious to the water.
Out from Independence moved the scattered parties of
traders to rendezvous at Council Grove some hundred
and fifty miles away on the Neosho, and there to organ-
ize their caravan under a captain. Apaches beset the
trail to murder and plunder, wolves to prey, and droves
of wild horses to stampede the live stock of the train.
The wagons were formed into a hollow square at night
as defense against the Indian or corral for the animals.
Every man had to take his turn at guard duty, according to " the common law of the prairies."

The outgoing caravans were to trade with the non-
industrial population of New Mexico; hence their cargoes included every manufactured commodity from salt
to silk. The exchanges included specie, gold and silver
bullion, gold-dust from the placer mine near Santa Fé,
buffalo rugs, furs from western trappers, some wool of
poor quality, a few coarse Mexican blankets which could
find a market in the American frontier settlements, and
immense droves of mules and horses.[15] Most of these
were products of a people lingering in the hunting and
pastoral stages of civilization. The value of the merchandise exported from the United States after 1826,
when wagons were regularly used for transportation,
averaged about $140,000 per annum, but it rose to
$250,000 in 1831 and 1839, and to $450,000 in 1843.
The goods were sold at a great advance, and the profits
were doubtless large, but the business was most hazard-
ous. Moreover, it was in the hands of a large number

of small dealers, who invested their all in the cargoes of
their wagons. The number of men in these yearly caravans was always large in proportion to the transac-
tions, indicating that the human intercourse between
the United States and New Mexico was more significant
than the commercial.

Soon Americans settled in Santa Fé, monopolized its
trade, and then extended their commercial operations
over a still greater field. Stretching away for three
hundred and twenty miles directly southward, the valley
of the Rio Grande, like a deep groove between the
flanking ranges, opened a way to Mexico through the
water-gap of El Paso in the mountains. This had
been the line of early Spanish expansion northward into
New Mexico and the San Luis Valley of Colorado, and
became now the route along which American commerce
spread to the southward. The Rio Grande is not navigable in this upper course, but the great national highway along its banks carried American traders with their
pack-horse trains to El Paso, whence they continued
two hundred miles south to Chihuahua, an important
administrative and commercial center of northern
Mexico. The trade with this place became particularly
active and profitable, so that between 1830 and 1840 it
absorbed nearly one half the imports of the Missouri
caravans.[16]

Thus American commercial expansion found a natural opening to a market in the south. To the west there
was no such point of attraction for the merchant, but
for the trapper, the wild, rugged uplands of the Rocky
Mountains, their snow-fed streams, and untouched game

preserves formed an alluring field of operations. The geographical location of Sante Fé made it a center from which in a wide sweep he could follow his favorite occupation. Therefore in the next advance to the west the trapper was again the pioneer. The trader keeps to a beaten road between two points, but it is the nature of the trapper to penetrate a whole river system, following its main course and the twig-like branches of its head streams, in order to set his traps for beaver or hunt deer in the isolated meadows of the upland valleys.

Santa Fé lies at an elevation of 7000 feet, on the north and south course of the Rio Grande about half-way between two points where this river is closely approached by two westward-flowing streams, the Gila and the Colorado, which themselves unite at the southeastern corner of California. In other words, the Colorado system, by means of the Gila to the south, the Grand River and Gunnison to the north, spans the valley of the Rio Grande from southern New Mexico to southern Colorado. In Santa Fé American energy had a foothold within the rim of the Rockies for further expansion. The well-beaten roads along the banks of the Rio Grande led in either direction to the narrow watersheds beyond which lay the rushing sources of the west-bound rivers. Hence we shall see that Santa Fé, in consequence of its geographical location, became the center of the first sustained expansion to the Pacific; that the two main routes followed were determined by the Gila and the upper Colorado; and that consequently southern California was the first portion of the Pacific coast to experience any regular overland intercourse with the United States.

From the Rio Grande in southwestern New Mexico to the sources of the Gila River is only about fifty miles. The watershed is formed by the Mimbres Mountains, which rise only three or four thousand feet above the Rio Grande, here flowing in a valley four thousand feet above the level of the sea. At a very early date there was an established trail over these mountains as far as the Copper Mines, which were worked by Spanish or New Mexican proprietors, or leased to Americans. They lay twenty miles northwest of the upper Mimbres River, about twenty miles east of the present town of Silver City at an elevation of over six thousand feet, and about thirty-five miles, or a two days' journey, from an eastward bend of the Gila or "Helay." Pack-mules bringing out the metal and carrying fuel into the mountains for the primitive smelting works had worn a distinct trace. This branched off from the main thoroughfare of the Rio Grande valley at 33° 20' north latitude near the present village of Palomas and led southwest a four days' journey to the mines. It ran through a beautiful grazing country, over streams flowing down from the Mimbres Mountains, ascended one of these narrow valleys, and crossed the range by a pass so easy that the summit was left behind several miles before the fact was discovered; then down the dry bed of an arroyo to the Rio Mimbres and the mines.[17] Just beyond, the countless branches of the Gila sources harbored beavers in considerable numbers.

All this region became a natural field of activity for the trappers from Santa Fé; and the direct western course of the Gila in no very long time guided them to the border of California. This was the shortest route across the continent from St. Louis. The marked south-eastward bend of the California coast from Cape Mendocino brings San Diego five degrees of longitude farther east than San Francisco, and within twelve degrees of old Fort Union in the Mora valley of the Canadian River. After the cession of all this southwestern country by Mexico in 1848, the first government surveys for a transcontinental railroad were made along the valley of the Gila.

The first overland party to enter California was that of Jedediah Smith in 1826, who took a more northern route from an advance trading-post on Great Salt Lake, southwest to San Diego. But his purpose was merely to explore and therefore his venture may be regarded as a sporadic effort, especially since it did not establish regular communication, as did the early advance from Santa Fé along the Gila route. The pioneers in this movement were the Patties, father and son, in 1827. The career of these men is typical. Natives of Kentucky, they became lumbermen in Missouri, in 1824 joined a trapping and trading expedition to New Mexico, and by the next three years they brought in skins from the Gila more than once. Then the elder Pattie organized a party of thirty trappers at Santa Fé to operate on the Colorado. Only eight of these reached the mouth of the Gila. Thence they floated down the Colorado in canoe-rafts and reached Santa Catalina mission

in Lower California in January, 1828.[18]   Ewing Young,
a Tennessean, came with a party of beaver-hunters from
New Mexico in 1830 to exploit the streams of southern
California, and again the next year, trapping on the
Gila as he went.   Jackson in 1831 left Santa Fé by
the southern route with nine hired men to buy mules
for the Louisiana country.[19]   Their route was traversed
by several parties in the next few years, for it was
always an open one; and in spite of the desert and
Apache attacks it was comparatively safe for even small
bands of travelers.   Some members of all these early
parties, after one or two trips, settled in California and
planted the first milestones, as it were, of American
expansion to the Pacific.

The Gila Trail formed the route of Kit Carson's
famous ride in 1846 from California, when he came
bearing dispatches from Commodore Stockton for the
authorities in Washington concerning the part of
the Americans in the Mexican War in California;
and it was the line of General Kearny's march the same
year to take charge of the military operations on the
Pacific coast.   The Trail was practicable for horses and
was well supplied with water, but further south the
Mimbres Mountains sink into the Sierra Madre Plateau,
which afforded a more arid though almost level route
to the West.   Skirting this southern rim of the Gila
basin, it was possible to pass westward at an elevation
of three or four thousand feet, and then drop down to
the regular Gila Trail just above the great bend.
Therefore Kearny sent his wagon-train under Lieutenant Cooke to follow by this more level path which,

once being opened, was utilized by Californian emigrants in subsequent years. Moreover, its value as an easy passage for a national railroad to the Pacific became apparent to W. H. Emory of the Mexican Boundary Commission, and was later a strong motive in the purchase by the United States of the territory south of the Gila.[20]

In the mean time, the restless activity of the trapper had pushed northward from Santa Fé to Albiquiu and along the valley of the little Chama River over the divide and westward across the sources of the San Juan to the Dolores and the Colorado; or these same untiring explorers had moved up the high valley of the Rio Grande and into the elevated basin (7500 to 8000 feet elevation) known as the San Luis Valley, from the northern end of which, by pass and stream-worn gorges, they traversed all the wild upheaved area of the central Rockies, and penetrated to all the upper tributaries of the Colorado, — the Gunnison, Grand, and Green rivers. The ranges and the passes were high, but so was the base of the trappers' operations. Taos, the center of trade and population on the upper Rio Grande, was about sixty-five hundred feet above the sea. The head streams of the Colorado system afforded fine hunting-fields, especially near their sources; for the cañon formation of the chief branches and the trunk streams impeded the ubiquitous searching of the trappers. Hence in their progress to more western hunting-grounds, these men of the woods passed along by the minor watersheds and small transverse valleys of the Colorado affluents, avoiding the main streams. When they had reached the inner rim of the Great Basin, the next obvious step was to the Pacific.

William Wolfskill of Kentucky, after trapping and trading extensively for eight years about Santa Fé, fitted out a party in 1830 to trap in California. He set out from Taos by a northerly route, swung around westward across the Grand and Green rivers and over the Wahsatch Mountains to the Sevier River. This is a stream of the Great Basin, flowing northward from the divide which the Virgin River, an affluent of the Colorado, drains to the south. Hence his path turned southward up the Sevier, over the mountain rim of the Basin, and southwest down the Virgin. The vast cañons of all these western rivers in their highland course preclude navigation even when the volume of water would permit it. Therefore Wolfskill turned away from the Virgin before reaching the Colorado, and struck out southwest across the wide Mohave Desert and reached the Cajon Pass (4560 feet) in the San Bernardino Range, through which he moved down to Los Angeles.[21]

This route came to be known as the Spanish Trail. It settled down to the more direct line up the Chama River and down the Dolores valley to the Grand. This it crossed near the present town of Moab, Utah, where a post-road now leads over the river, and took advantage of a break in the cañon formation of the Green to pass that stream just where the railroad crosses it to-day.[22] In time it was well defined, for along it moved yearly caravans from Santa Fé to California led by Americans or New Mexicans, bringing woolen fabrics and zarapes to exchange for the mules and horses of California or the silks and other choice imports of China.[23] It was the route followed by Fremont on his return from California in 1844. The Mohave section, like all desert trails, zigzagged across the arid waste from one watering-place to another. These were springs, water-holes, or little wells, dug narrow and deep in the dry bed of the Mohave River by wolves, whose keen scent told them of the underground water supply.[24] Sometimes there were stretches of forty or sixty miles without water. It was impossible to miss the Trail here, because it was traced with bones. After a twelve or fifteen hours march without relief, the animals, exhausted by the scorching heat and parched with thirst, would stagger along with drooping head and tail, when suddenly the wild mules which generally accompanied the caravans as return products or reserve pack-animals, would toss their heads with an alert movement, sniff the air, and then start off at a full gallop down the dusty pathway. This brought joy to the men of the Trail, because they knew there was a surface supply of water a mile or two ahead.

Such were the two routes, geographically determined, from New Mexico to California. Both had their natural termini in San Diego and Los Angeles. The men who came to this country to trap continued their operations along the river feeders of Tulare Lake and the mountain tributaries of the San Joaquin, and lived the nomadic lives of hunters. Though a steady intercourse was kept up by the annual visits of the Santa Fé traders, these southern routes never became avenues of heavy immigration. The trapper and the trader are not domestic animals; they bring no families with them except mongrel offspring of Indian wives. Mountain and desert acted as a barrier, and the overland infiltration of American elements into California in early days went on very slowly. Only thirty hunters were added to its population between 1831 and 1835, the residuum of the Santa Fé parties and of two other expeditions which came directly across the Sierra Nevada to the Pacific.[25] More Americans came by sea. The New England trading-vessels, which yearly visited this coast in goodly numbers, added more to the American colony, for they were constantly leaving here invalided sailors, deserters, and occasionally a commercial agent. So an American population in California was slowly being recruited, exclusively men who in general married Californian women. The steady influx which later planted a bit of the United States on California soil was intimately connected with the peopling of the contiguous territory of Oregon, where the United States frontier met that of California.

NOTES TO CHAPTER X

1. Elliott Coues, History of Lewis and Clark Expedition, vol. i. p. 150. 1893.
2. Ibid. pp. 178, 203.
3. Ibid. p. 273.
4. Ibid. vol. ii. p. 1116.
5. Ibid. vol. ii. p. 1206.
6. Fremont's Narrative, p. 17, Washington, 1845, and Parkman, Oregon Trail, pp. 87, 88. 1901.
7. Irving, Astoria, pp. 134, 178. 1854.
8. Parkman, Oregon Trail, chap. ix.

9. Fremont's Narrative, pp. 35–95. 1845.

   H. M. Chittenden, The American Fur Trade of the Far West, vol. iii. pp. 947–970, and map. New York, 1902.

10. Josiah Gregg, The Commerce of the Prairies, vol. i. pp. 32, 33. 1845.

    Inman, the Old Santa Fé Trail, pp. 142–145. 1897.

11. Gregg, The Commerce of the Prairies, vol. i. pp. 111 and 145. 1845.

12. Ibid. vol. i. pp. 18–21.

    E. Coues, Pike Expedition, vol. ii. pp. 490–493. 1895.

13. Gregg, The Commerce of the Prairies, vol. ii. p. 155.

14. Ibid. vol. i. pp. 108, 109.

15. Ibid. vol. i. p. 307.

16. Ibid. vol. ii. pp. 161–164.

17. W. H. Emory, Notes of a Military Reconnoissance from Fort Leavenworth in Missouri to San Diego in California. Made in 1846–47. Map and pp. 55–59. Ex. Doc. no. 41. Washington, 1848.

    Personal Narrative of James O. Pattie of Kentucky, p. 52. Cincinnati, 1831.

18. Ibid. pp. 51–57, 133–137.

    Bancroft, History of California, vol. iii. pp. 162, 163. 1884.

19. Ibid. pp. 174, 387.

20. W. H. Emory, Report of the United States and Mexico Boundary Commission, vol. i. p. 93, 1857.

    Bancroft, Arizona and New Mexico, pp. 477, 478. 1889.

    Lieutenant-Colonel Cooke, Report of March from Sante Fé, New Mexico, to San Diego, California, pp. 553–557. Ex. Doc. no. 41. Washington, 1848.

21. Bancroft, History of California, vol. iii. p. 386.

22. Senate Report of Explorations and Surveys for a Railroad Route from the Mississippi River to the Pacific, vol. xi. part ii. Map 4, by Captain E. G. Beckwith. Washington, 1861.

    Lieutenant Abert, Examination of New Mexico, 1846–47. Ex. Doc no. 41. Washington, 1848.

23. Bancroft, History of California, vol. iii. p. 395.

24. Fremont's Narrative, p. 245. 1845.

25. Bancroft, History of California, vol. iii. p. 373.

CHAPTER XI

EXPANSION INTO THE FAR WEST BY THE NORTHERN TRAILS

AT the time the American trappers and traders found their way to the Pacific, California was a foreign possession. This fact, reinforced by the length and difficulty of the journey thither, sufficed to discourage the immigration of families, especially after the rebellion of Texas under American leadership had rendered citizens of the United States undesirable tenants of Mexican soil. Texas, to be sure, had been Mexican domain, but it was close to Louisiana and the Mississippi River. Oregon was remote as California, and like it was barred by two thousand miles of plains, mountains, and desert; but it was claimed on solid grounds by the United States. Hence in this direction Americans turned when the uneasy spirit of migration began to stir along the Missouri and Mississippi frontier.

The mouth of the Columbia had been discovered by a New England vessel in 1792; it had been reached by the overland exploring expedition of Lewis and Clark in 1805, and occupied temporarily as a trading-point by the fort of the Astor party in 1811. The line of the Missouri and Columbia first opened the way to the Oregon country; immigration set vigorously into this trans-Rocky America, and from the southward-pointing valley of the Willamette spread over into the vast trough of northern California. Thus California was entered freely at its northern and southern extremities by natural avenues, while on the east the double barrier of the snow-capped Sierras and the vast expanse of the Nevada Desert long excluded immigration.

From the point where the Colorado, after a long cañon-bound course, pours its thin, desert-worn current into the Gulf of California, northward to Vancouver Island there is only one river which traverses the whole width of the trans-Rocky region and breaks through the double mountain wall to the sea. This is the Columbia, claimed by the United States and counterclaimed by the British; for both it was the key to western expansion on the Pacific. Its three long arms penetrate to the heart and even to the eastern rim of the Cordilleras; they stretch a mighty span from the head waters of the Athabasca in the north to the sources of the Platte and the Colorado in the south, touching fingers along the Great Divide with the divergent affluents of the Missouri. These Columbia streams, with their slender supply of water and turbulent current, would rarely hear the dip of paddle or ripple of canoe-prow upon their cañon-shadowed surface; but they guided the trail down those rocky slopes from where the mountains doffed their caps of snow and flung wide their giant portals to the march of an invading people.

The Columbia was first approached from the east through the Missouri, where it takes its rise in the western ranges of the Rockies. The Lewis and Clark expedition therefore followed this river to its forks in the heart of the mountains, continued up the Jefferson, its most westward branch, to the outermost range, whence the Lemhi Pass and an Indian trail led down to the Lemhi branch of the Salmon River.[1] Hopeless of traversing the wide desert by the mad, rock-bound waters of the Salmon, the party turned northward over the range and up the Clark River, which flows between the parallel ranges of the Rockies.   Its lateral affluents, corrading transverse valleys, served as intermediary paths over the mountains between the upper Missouri and the central tributaries of the Columbia, just as farther south the Green River branch of the Colorado by its side valleys opened a way over two watersheds for the Oregon Trail from the Sweetwater branch of the North Platte to the Port Neuf and the Snake River at Fort Hall. Such was the part played by the New and the Holston in the early trails across the Alleghenies ; and earlier still, such was the anthropo-geographical importance in the Alps of the torrent-scoured gorges branching off from the valleys of the Upper Rhine, Rhone, and Inn in the lines of communication between Italy and the northern plains of Switzerland.[2]

The return route of Captain Clark led directly east from the great elbow of the Columbia up the Snake and Clearwater River, over the Bitter Root range by the Lolo Pass, and by the Lou Lou Fork and Blackfoot valley of the Clark River to the Lewis and Clark Pass, down to the little Dearborn River of the Missouri. A well-beaten road marked the whole trail, for it was the path of the western Indians coming to hunt buffalo on

the plains.[3] This was the shortest route across the
mountains, because the Rockies contract as they near
Canada; this route had the longest river approach on
a navigable stream, the best mountain pass, the narrowest stretch of desert on the west, and the most direct
approach to the great bend of the Columbia, where
regular canoe navigation began; but it involved too
great a detour to the north. It was too far from St.
Louis and that bustling center of emigrant preparation
at the elbow of the Missouri. Hence we find that
the Astor expedition, which set out in 1810 to establish an American trading-station at the mouth of the
Columbia, went up the Missouri only to the Grand
near the Aricara villages, and then made a short cut
by striking out southwest across the plains to the Big
Horn branch of the Yellowstone; thence it continued
up the Wind River to the great central dome (9000 ft.)
of the Rockies, defined by the Teton, Gros Ventre,
and Shoshone ranges. This it crossed by a single pass,
analogous to the St. Gotthard in the Alps system, to a
head stream of the Snake River.

A returning party of Astoria traders a year or two
later struck southeast from the upper Snake, and reaching the Green River basin, took advantage of the general dip in this section of the Rockies to reach the
Sweetwater, the North Platte, and finally St. Louis.[4]
This was in its main features the route followed later
by the Oregon Trail when expansion to the Pacific
began in earnest. The Astoria venture was short-lived
because it had to contend with an overland line of
communication hopelessly long and difficult, without

any base between the Mississippi and the Pacific; with interruptions to its sea communications caused by the War of 1812; and with territorial encroachments of the British in the aggressive expansion of the Hudson Bay Company into the valley of the Columbia.

The claim of the British to the Columbia was geographically based. The east and west line of the Saskatchewan River had at a very early date carried English explorers in Canada to the northern arm of the Columbia among the Selkirk Rockies. Discovery of the mouth of the river, by an international principle which the English themselves had established, gave this great stream to the United States; but its northern source was in the hands of the British. We have seen how easy and natural is downstream expansion. The great trading company had a near base in their fur stations on the upper waters of the Canadian rivers. Wealth, organization, a long-established service, and strong political backing gave them an effectiveness which Astor's American Fur Company lacked. Aided by the happy accident of war, they were able to extinguish by the seizure of Astoria the last spark of life in the American Fur Company on the Columbia. Then they proceeded to take systematic possession of the whole Columbia basin after the manner of the fur-trader which the French in Canada had found effective for a time.

By 1834 every strategic point at bend or fork of stream, and every center of the great fur-hunting fields in the Rockies was marked by a fortified post. Fort George, on the site of the old Astoria, maintained communications with the sea. Fort Vancouver, the chief administrative post, six miles above the mouth of the Willamette River, exploited the rich farm-lands of that valley to supply the upstream line of forts with bread. Where the wide diverging arms of the Columbia met at the great bend, Fort Walla Walla held a position as vital for the control of this sole western waterway as old Fort Duquesne for the command of the Ohio. It felt the pulse of the wild trapping life from Fort Hall on the upper Snake, drawing in its beaver from the slopes of the Wahsatch Mountains, to Fort Colville, far up the Columbia at the foot of the Selkirk Rockies.[5]

As we have seen before, a primeval wilderness, widely separated trading-posts, a sparse population of nomadic trappers half assimilated to the savage life about them, and above all a monopoly, belong to the nature of the fur trade. The influence of the Hudson Bay Company was sufficient to keep out the English traders and settlers from this great game preserve, and they could undersell the American traders;[6] but they could not exclude American settlers under the agreement of joint occupation made with the United States in 1818. The experience of the French had proved how weak a land tenure was that of the trading-post when contesting claims with a sedentary agricultural population. At first it was only a few American trappers who sifted through the mountain breaches into the Oregon country; but soon a line of dust hovering in the air marked from afar the Oregon Trail. Into the wide gaps between the stations and forts poured an inundating tide of American immigrants, sweeping away the beaver and the game, setting up the claim of plough and saw-mill against the weaker title of the trap.

The Hudson Bay Company, through the natural avenues of the Salmon and Snake rivers, made its power felt as far as Great Salt Lake, the Bear River, and the Green River basin; but here it came into conflict with American trappers who entered this famous fur-field from every direction, from the upper Rio Grande, Arkansas, North Platte and Sweetwater, and the Wind River branch of the Big Horn. Rivalry was fierce. The American and the Rocky Mountain Fur Company had their traders' rendezvous sometimes on the Wind River, more often on the Green, and sometimes even at Pierre's Hole,[7] a valley just west of the Teton Range and draining into the Henry branch of the Snake; but occasionally American traders pushed into the British rendezvous at Ogden's Hole on Bear River within the Great Basin. All this area of interlocking streams was the battle-ground of rival commercial interests; but though the American got a slight foothold on the westward-flowing streams, encroachment was limited in general to the sources. The Americans had an advantage to the southwest. Ashley of the Rocky Mountain Fur Company made his first rendezvous on the Green River in 1824, entering its valley by the famous South Pass; and the next year founded Fort Ashley on Utah Lake, where he left a hundred men to exploit the peltries of this rich country.[8]

Between 1826 and 1829 there were about six hundred American trappers in the region of the Teton

Mountains and Green River, also a great number of the Hudson Bay Company men.[9] The wild sons of the trail explored every upland valley and pass, and opened the mountain paths for others, but there is little evidence that they contributed to the settled population of Oregon until after the dissolution of the American Fur Company in 1840. Furthermore, it was comparatively late before a regular westward movement of population set towards Oregon.

The first immigration into that country was the result of individual enterprise originating on the Atlantic seaboard, and having nothing in common with a national expansion of the frontier. It was recruited from a population which had left its pioneer days far behind and therefore had unlearned the useful lessons of the wilderness; or it drew upon Canada's ready-made stock of traders and trappers. It came in no small part by sea and relied upon maritime connection with the East for supplies, support, and growth. Such was the Astoria party, recruited almost entirely both as to partners, factors, and engagés from the Scotchmen and French Canadians of Montreal and Mackinac. Such was the Massachusetts expedition under Wyeth, which came overland in 1832 to establish a trading-post at the mouth of the Columbia. The journey of over a thousand miles from St. Louis to the Snake River in the company of the veteran Sublette and his Rocky Mountain trappers educated the inexperienced but purposeful Wyeth to the most important pioneer requirements. Some of his fantastic equipments, elaborately devised in the scholarly environment of Cambridge, were laughed

out of service by the practical men of the plains at the
Missouri rendezvous; the rest were exhausted or lost, or
discarded when the strength of himself or his pack-
animals began to fail, so that his diminished party
reached the Columbia almost empty-handed.   The vessel which was to bring merchandise around the Horn
for his post never arrived; and though a return to the
East and reinforcements of his supplies justified him in
building his stations on Wapato Island in the lower
Columbia and at Fort Hall on the Snake, the Hudson
Bay Company was too strongly entrenched in all the
Oregon country to suffer this invasion of its trading-
grounds.[10]

While the men who knew the ground, the American
trappers and traders, were still hanging on the mountain outskirts of the Oregon country, religious enthusi-
asts in the eastern states were fitting out missionary
parties for the uplifting of the western Indian, their
zeal being as usual in direct proportion to their igno-
rance.   Such was the beginning in 1834 of the Metho-
dist mission in the valley of the Willamette and the
Presbyterian stations on the Walla Walla and the Clearwater branch of the Snake River in 1836.   The pioneers
came overland by the North Platte and the Snake River.
The inland location of the Presbyterian missions, the
poverty of the country about, and the dangerous character of the Indians, though at the river terminus of the
desert trail, prevented their attracting reinforcements
and growing to a settlement.   The Willamette mission,
accessible to the mouth of the Columbia, drew from
Boston and the East by sea the supplies and people

which the wide expanse of plain and mountain and desert at that period debarred.[11]

Jason Lee, finding the Indians of the Willamette impossible and the country highly possible, was converted from missionary to colonizer. His lectures in the United States advocating the settlement and formal appropriation of this western possession before the clutch of the British should be tightened upon it, writings in similar strain of other pioneers who had tried to get a foothold in the Columbia trade, and the discussion of the " Oregon Question " in Congress, served to advertise the trans-Rocky country. Lee, on a return trip to the United States, met his first audience and the first response to his eloquence in the frontier states. In the spring of 1839 two small parties moved out from Illinois on the route to the Columbia,[12] the forerunners of a new westward movement. Here was a genuine expansion from our western frontier, inaugurating the second phase of Oregon immigration.

Up to this time the settlements in Oregon were an anachronism, — our farthest West with the stamp of our farthest East ; social, political, and economic ideals of New England storekeeper, mill-owner, preacher, and school-teacher in the vast Cascade forests and the wilderness valleys of the Willamette and Columbia. But the eastern settlers were too few and the Atlantic seaboard too remote for this to last. As the irregular explorations of the West reduced the Oregon Trail to its easiest and final route, the Missouri frontier was brought nearer. By 1840 the rapid occupation of the early western states, the lack of other outlets than

the Mississippi and New Orleans, and the consequent
gorging of that market without relief for the plethora,
caused a disastrous drop in the prices of western agricultural products. One Missouri farmer sold a boat-load
of bacon and lard for a hundred dollars. The Mississippi steamboats at times found in bacon a hot and
cheap fuel. So much for living in a country without a
seacoast. And the western farmer, with the migratory
instinct in his blood, set out without regret to the fertile
plains of the Oregon country, whence the wide Pacific
highway led to Asiatic markets.

From that time the dust of travel thickened on the
Oregon Trail. From Arkansas, Missouri, and Iowa, from
Kentucky, Tennessee, and Illinois [13] the pioneers gathered
at the rendezvous at Independence, Liberty, St. Joseph,
and Council Bluffs, with their heavy moving-wagons and
their lowing herds. The Trail, 2400 miles long, led up
the Platte and its North Fork to Fort Laramie, around
the Black Mountains by the Sweetwater to South Pass
(7490 ft. elevation), over this dip to the Green River, up
the transverse valley of the Black River, by Muddy Creek
to a pass over the divide rimming the Bear River valley
on the east. This was the highest point of the whole
long trail, 8230 feet elevation.[14] The road dropped
down the Bear River to the most northern point of its
course and then crossed over an insignificant watershed
to Port Neuf River and Fort Hall on the Snake. Thus
far the way was easy. To South Pass the ascent was so
gradual that it was difficult to determine when the summit was reached, and the grassy road to the higher pass
over the Bear River divide was only " at times steeper

than the national road in the Alleghanies." [15] The
pass itself was a good one, and the fertile meadows of
the Bear River valley were a natural recruiting-ground
for the travel-worn caravans.

Fort Hall was 1200 miles from Independence and only
a little over halfway to the mouth of the Columbia.
But the last part of the road was the hardest. For three
hundred miles across the desert without a fertile spot or
any pasturage, the trail followed the Snake River, whose
cañon walls for days together barred the thirsting herds
from its rushing waters. From Salmon Falls the road,
avoiding a wide bend of the river, cut across the plains
to Fort Boise, an intermediate station of the Hudson
Bay Company, whence it continued northward down the
Snake again to Burnt River. Now only the thickly
wooded range of the Blue Mountains lay between the
emigrant and the Columbia. The road turned off up
the Burnt River Cañon, over a dividing ridge to the
upper Powder River, whose transverse valley pointed
the line of easiest ascent up the steep slope. The range
once surmounted by double teams, the Umatilla opened
an easy path down to the Walla Walla and the great
bend of the Columbia. [16]

Four months of continuous traveling, over two thousand miles of weary plodding over plain, mountain,
desert, and mountain again, with two hundred and
fifty miles of dangerous navigation on the rapid-swept
Columbia; days and weeks of scanty food and scantier
drink, of midday heat and midnight cold in the plateau
desert, — these were the tests which geographical conditions set to the survival of the fittest on the Oregon

Trail. Faint-hearted members of the caravans turned back at Fort Laramie, Green River, or even Fort Hall. In the earlier days wagons were exchanged for pack-horses at Fort Laramie; but as the road became better known, the lumbering vehicles were carried farther, — to Fort Hall and finally to the Columbia. Property was abandoned all along the route as pack or draught animals dropped from exhaustion or were used for food.[17] Even the Columbia did not bring relief. The lean and sore-footed herds, the hope of future ranches and the leather trade with China over the wide highway of the Pacific, might reach its banks, but they could not be transported in the frail canoes and bateaux which shot its rapids. The only alternative was to sell them at a sacrifice to the British at Fort Walla Walla, or to start out on another weary journey over the wall of the Cascade Mountains along a dangerous trail beset by thieving Indians who carried off prizes from every herd.

But in spite of all these obstacles, on they came by the hundred and the thousand, a sturdy western stock, large in the patience and hospitality born of their experience, broad in their democracy, insistent upon their rights, aggressive and even blustering in their Americanism. They settled in the valleys of the Willamette and its tributaries, on the Nehalem River flowing into the Pacific, the plains about Astoria, and in spite of British protests staked their claims about the head of Puget Sound.[18] Along the Willamette they found a mixed population. In the ascendency were the original mission settlers from New York and New England, who in 1843 had been reinforced by small parties from

Massachusetts and Maine, coming by sea and well provided with worldly goods. These had come to set up
stores, build mills, and speculate in land.[19] Living among
the men from the East but holding themselves aloof were
a considerable number of French Canadians, old employees of Fort Vancouver, who now raised wheat for the
post.[20] These, too, had received accessions from an
unexpected source. In 1841 the Hudson Bay Company,
in order to strengthen its hold on the Oregon country,
brought out a colony of sixty people from the Red River
settlement of Canada by the way of the Saskatchewan
to take up land on Puget Sound about the Nisqually
River. But the colonists, finding the soil poor, removed to the Willamette valley, thus defeating the
political design of the British and strengthening the
American center.[21]

But now the Willamette settlement was inundated by
the inpouring tide from the Missouri frontier. The
new immigrants by their number were able to remove
from Oregon the local stamp of the East and of the
Hudson Bay Company, and to put upon its political and
social institutions the stamp of a larger Americanism.
"They carried the world of politics on their shoulders."
The petty provisional government set up under missionary domination in 1843 they fundamentally altered
in 1844 to conform to their own more democratic
ideals.

The presence of eight thousand Americans in the
Columbia country in 1845[22] hastened the settlement of
the "Oregon Question" in 1846. The appeals of the
Pacific colony for incorporation in the Union were ener-

getically supported by the western states, who from 1843 to 1846 assailed the Senate and Congress with resolutions from general assemblies, memorials, and petitions for the political occupation of Oregon. The bond of union between the Columbia and the Missouri frontier was close, for the geographical environment of the Mississippi valley had bred the expanding tendencies which peopled the valley of the Willamette.

As the most accessible and choicest land in Oregon was taken up, an open, direct road made by nature up the valley of the Willamette and down the long trough of the Sacramento invited dissatisfied colonists to seek more desirable fields in northern California. This was already a well-beaten trail. Before the coming of the missionary, the American trapper with well-laden pack-mule had taken this way up from California to the ready fur market at the Hudson Bay post of Vancouver.[23] The California cattle which stocked the Willamette ranches in the early days helped to beat out this trail,[24] and at all times there was more or less communication. Oregon received a few stray immigrants from California, and California got some of its earliest pioneers, like Sutter, from Oregon; but from 1843 to 1846 every year saw a large party from the Willamette cross the watershed to the Sacramento.

Nor was this the only expansion into northern California. Russians and English had already made a commercial invasion of the country from the relatively near bases on the Columbia River and Baranof Island. The Hudson Bay Company had a post on San Francisco Bay, whence its trappers ranged the upland valleys

of the Sierra Nevada.   Communication with Fort Vancouver was maintained both by land and sea.   The
Russian expansion was maritime; it was therefore easy
and appeared early.   In 1812 it built its trading-post
at Ross near Bodega Bay a few miles north of the
Golden Gate, and established a sub-station on the Farra-
lone Islands, where from twelve to fifteen hundred fur
seals were taken annually for the first few years by the
Aleutian hunters in their skin boats or *bidarkas*.   The
Russians wanted not only furs, whether of sea or land,
but also grain and other food supplies which they purchased from the Spanish Californians for their far
northern trading-stations on the Alaskan coast.[25]

Lake Superior, Lake Winnipeg, the Saskatchewan,
Columbia, Willamette, Sacramento had been the British
line of approach to San Francisco Bay; the more open
highway of the Pacific along "Russian America" and
its islands, that of the Slav advance; the Gila River,
Spanish Trail, and Oregon Trail, that of the American.
The presence of all these foreign elements in this northern province of Mexico points to the broad zone of
indefinite frontier characteristic of every new land.
Western boundaries were not yet evolved and western
territories had not yet been differentiated out of their
"indefinite, incoherent homogeneity."   The Americans,
English, and Russians were all regarded with some
apprehension by the Californians, though the advantages
of foreign trade overcame their misgivings.   It was the
United States which on the basis of an as yet unformu-
lated Monroe Doctrine called a halt to Russian expansion on this coast in 1823, when the line of 54° 40′

north latitude was agreed upon as the southern limit of Muscovite dominion in America.

Meanwhile the American element in this extra-United States territory was increasing. The efficacy of informal expansion in Texas was an argument to the practical mind of the American for the application of the same method to California. The public sentiment as to "manifest destiny" was pretty well formulated as early as 1841. Scouting parties of early immigrants, as we have seen, had found the northern and southern breaches in California's mountain wall; but now the advancing army began to cross the vast sandy moat of the Nevada Desert and assail the mighty rampart of the Sierras.

From the time that Ashley and Bonneville [26] had discovered Great Salt Lake and trading-posts had been established there, and on Utah Lake to the south, small bodies of trappers and explorers had been guided by the Humboldt River westward across the desert, and with their slender equipment had managed to scramble over the mountains into the great Valley of California. [27] They discovered many high-laid passes at the heads of the deep-cutting torrents which tore down the steep slope to be sucked up by the bibulous sand below, but none were practicable for a heavy-laden caravan of immigrants. As the accounts of California aroused greater interest in the states, an adventurous body of immigrants under Bidwell and Bartleson in 1841 struck off from the Oregon Trail at the northern bend of the Bear River, followed the course of that stream southward almost to Great Salt

Lake, and then turned west, searching for the Mary or Humboldt River, which they found after nearly a month and followed westward to Humboldt Sink. Thence they continued south to the Walker River, by means of its valley struggled up the mountains, crossed the ridge near Sonora Pass (10,115 feet elevation) with untold suffering from cold and starvation, and reached the San Joaquin valley by the Stanislaus River.[28]

The best lands in Oregon were now taken up, and the overflow into California was increasing; hence there were renewed efforts to find a more direct route to the valley of the Sacramento without making the long detour by way of Oregon or Los Angeles. Finally the Truckee Pass in the Sierras was discovered in 1844 and the California Trail settled down to the line of least resistance. It branched off from the Oregon Trail and the Snake River about a hundred miles below Fort Hall, turned southwest up Goose Creek, a southern affluent of the Snake, and over to the head waters of the Humboldt River. Later it left the Oregon Trail at Bear River. From Humboldt Sink the road led directly west to the Truckee River, up this stream to Truckee Pass (7017 feet) and down the western slope by the American or Bear River of California to the Sacramento valley. Sutter's Fort near the mouth of the Bear River was a distributing-point for immigrants entering both by the Sacramento route from Oregon and by the California Trail. Therefore this adobe settlement with its various industries formed a caravanserai, ever filling and emptying its contents into the Sacramento, Napa, and Sonoma valleys. With the opening of the California Trail, the immigrant parties increased in size and in 1845 began to include families.[29] Northern California was being Americanized and Sutter's Fort was the rallying-place. This trading-post on the frontier, remote from Mexican control emanating from Monterey or Los Angeles, accessible to the steady stream of Americans from the east and north, was in a position to "contribute materially to hasten the success of American occupation."[30]

But California was much farther from the United States frontier than Mexican Texas had been. Both the length and natural difficulties of the overland journey precluded immigration on the large scale that had made an Anglo-Saxon state out of a Latin province on the Sabine and Brazos. At the end of 1845 the total foreign male population of California was only 680;[31] but the majority of these were Americans and were collected around and about the Sacramento valley and San Francisco Bay.

But if California was far from the United States, it was almost as remote from Mexico, and the Mexicans on the ground not numerically strong ; 6900 for the whole department, 3550 for northern California.[32] In relation to the administrative center in the City of Mexico, California was distinctly a peripheral state with the usual tendencies towards defection. Conditions in the northern department were not understood in the far-away capital. Mexican governors and officials were unsatisfactory. Hence in 1836 and again in 1844 a successful revolution established the principle of home rule under national allegiance as against centralism, while there were advocates of entire independence.[33]

Officials in California ignored instructions for the control of that department emanating from the capital; and the federal authorities at the City of Mexico found the efficiency of their control sadly diminished by poverty, which complicated the difficulty of maintaining troops in the remote province. On many subjects the Californian and the Mexican point of view differed fundamentally. When in 1843 American immigration into California began to increase noticeably, Mexican orders were received excluding such newcomers, prohibiting retail trade by foreigners throughout the republic, and reserving the coasting-trade to Mexican subjects; but no attempt was made to enforce these decrees in California. On the contrary, American immigrants were well received, obtained land grants readily; and for the protection of Boston trading-vessels on the coast, the governor of California levied a duty on all foreign goods coming in from Mexican ports.[34] The long stretch weakened the arm of authority.

The federal sentiment was strongest in the south. San Diego fought for centralism in the revolution of 1836, and Los Angeles was the bulwark of Mexican allegiance against the revolutionary agitations of Monterey and the north.[35] It was southern California which made the only obstinate resistance to American conquest in 1846. This portion of the state had a closer sea connection with the national port of San Blas through the Gulf of California and the Pacific than that enjoyed by Monterey and San Francisco; and it was the natural terminus both of the Gila River Trail uniting it with Santa Fé and Chihuahua, and also of a more southern route leading through the province of Sonora in northern Mexico.

The California of the Sacramento valley and San Francisco Bay, on the other hand, was more isolated from the central authority and more open to American influences coming in by the two northern trails. Unlike the Russian and British traders, the Americans were settled on farms and ranches, had largely monopolized the commerce of the country, and had taken to themselves Californian wives; for until the Truckee Pass alleviated the worst hardships of the California Trail, there were few women among the immigrants. Bonds of property, commerce, and marriage united the American settlers to the Californians, who fully appreciated the advantage to the community of such enterprising elements. A certain assimilation of sentiments undoubtedly followed. When in 1846 the war between the United States and Mexico became imminent, many Californians foresaw the economic benefits which would fall to their country with an American régime.[36] General Vallejo, guardian of the northern frontier and the sturdiest character among Mexican subjects in this region, favored the annexation of California to the United States. Everywhere in the north the spirit of resistance to the threatened invasion was lukewarm at best.

The United States had been turning covetous eyes towards California since 1835. It needed ports along this mountain-bound coast. Floyd of Virginia in 1822 advocated the occupation of Oregon chiefly to secure the mouth of the Columbia as a harbor for American

whaling-vessels. In Jackson's administration an offer
was made to purchase that part of California between
37° and 42° north latitude in order to secure San Francisco Bay. A few years later so radical a measure was
no longer contemplated. West Florida and Texas had
taught the Americans that the expansion of the trader's
pack-horse and the rancher's herds was the surest
pledge of territorial aggrandizement. They were possessed therefore with a large patience, in spite of elo-
quent appeals for acquisition from lecture platform and
press, while immigration into California went steadily
on and did its work. Thus the question of annexation
was taking care of itself. Time was the only thing
necessary, and the Americans felt they had all the time
required.

But the United States government was not napping.
The Russians had withdrawn from the field. The
thinly scattered trappers and isolated posts of the Hudson Bay Company in California could not vie with the
spreading farms and growing families of the American
settlers, but Englishmen who had large claims against
the Mexican government for money loans had been for
some time negotiating for a cession of territory in
California or elsewhere as payment or security for the
same.[37] The English government took no part in this
so far as known. Any effort on the part of any foreign power to seize any part of California was to be
checked on the basis of the Monroe Doctrine. This
was always held in reserve. Towards California the
greatest diplomacy was shown. If that province should

declare its independence of Mexico, it should be encouraged and supported as a "sister republic," and later invited to join the Union.[38] American settlers and American influence were to be relied upon to dictate the answer.

The American element in California, however, was less patient than the government. When rumors of a war between the United States and Mexico became louder, these men, remote from succor, threatened as some believed with expulsion from their hard-won homes, outnumbered but not outmatched by the Californians, acted upon the instincts of self-help bred on the plains and the frontier. They raised the flag of revolt in the Sacramento, Napa, and Sonoma valleys; and the American colors were flying over Sutter's Fort and the captured Sonoma when Sloat under American orders hauled down the Mexican flag at Monterey on July 11, 1846. Filibuster venture though it was, and complicate though it did the American occupation of the country by antagonizing the Californians, it strengthened Sloat's position on the coast and supported Kearny's troops when they appeared in southern California after their forced march westward along the Gila Trail.

In the treaty of 1848 the United States forced Mexico to yield up not only New Mexico and California, which American troops had conquered, but all the intervening country from the Rio Grande to the Pacific and from the Gila north to the boundary of Oregon. The lines of communication with the new

territory had to be secured.   The United States now
realized "the manifest destiny" of the American people to occupy the continent from ocean to ocean.

NOTES TO CHAPTER XI

1. Elliott Coues, Lewis and Clark Expedition, vol. ii. pp. 484–493.
2. Ratzel, Politische Geographie, pp. 684 and 688.
   E. C. Semple, Mountain Passes.   Bul. Amer. Geog. Soc. vol. xxxiii.
   nos. 2 and 3.   1901.
3. Elliott Coues, Lewis and Clark Expedition, vol. ii. pp. 1060–1070.
4. Irving, Astoria, chaps. 43–50.
5. Bancroft, History of Oregon, vol. i. pp. 6–14.   1886.
6. Ibid. vol. i. pp. 48 and 70.
7. H. M. Chittenden, The American Fur Trade of the Far West, vol. i.
   pp. 296–299.   1902.
   Bancroft, History of the Northwest Coast, vol. ii. pp. 455 and 562.
8. Ibid. vol. ii. p. 448.
9. Ibid. vol. ii. p. 459.
10. Ibid. vol. ii. pp. 557–564.
11. Bancroft, History of Oregon, vol. i. pp. 78–182.
12. Ibid. vol. i. pp. 227–237.
13. Ibid. vol. i. p. 393.
14. Fremont's Narrative, pp. 112–116.   1845.
15. Ibid. p. 60.
16. Bancroft, History of Oregon, vol. i. pp. 400–402, and Fremont's
    Narrative, pp. 147–166.
17. Parkman, The Oregon Trail, pp. 101–103.   1901.
18. Bancroft, History of Oregon, vol. i. pp. 413–415, 458–464.
19. Ibid. vol. i. pp. 422–424.
20. Ibid. vol. i. p. 15.
21. Ibid. vol. i. p. 252.
22. Monette, History of the Valley of the Mississippi, vol. i. p. 569,
    footnote.   1846.
23. Bancroft, History of the Northwest Coast, vol. ii. p. 455.
24. Bancroft, History of Oregon, vol. i. chap. vi.
25. Bancroft, History of California, vol. i. pp. 298, 628–635.
26. Irving's Bonneville.
27. Bancroft, History of the Northwest Coast, vol. ii. p. 449, and History
    of California, vol. iii. pp. 389–391.

28. Bancroft, History of California, vol. iv. pp. 269–271.
29. Ibid. vol. iv. p. 578.
30. Ibid. vol. iv. p. 227.
31. Ibid. vol. iv. p. 588.
32. Ibid. vol. iv. p. 649.
33. Ibid. vol. iii. p. 468–475.
34. Ibid. vol. iv. pp. 380–385, 429, 555.
35. Ibid. vol. iii. pp. 613–615, 629.
36. Ibid. vol. v. pp. 73 and 113.
37. Ibid. vol. iv. p. 298.
38. Ibid. vol. v. pp. 74 and 196.

CHAPTER XII

GROWTH OF THE UNITED STATES TO A CONTINENTAL POWER GEOGRAPHICALLY DETERMINED

From a narrow strip on the Atlantic littoral, the United States had pushed steadily westward until they had possessed themselves of the most desirable stretch of coast on the Pacific. In this advance from ocean to ocean geographic conditions, in the cumulative effects of their direct and indirect operation, became factors so strong that just for the sturdy energy of the Anglo-Saxon race they became determinants. A less vigorous people would hardly have responded to the educative influences of this peculiar environment.

The remoteness of the North American continent from the centers of civilization and teeming populations of Europe and Asia had maintained it as a great territorial reserve, occupied only by a sparse population whom the want of propitious geographical environment had helped to keep in savagery or the lowest stages of barbarism.[1] Here were lacking those small segregated geographical areas which early curb the wandering instincts of a primitive people, bring them to a sedentary life, and then by protecting mountains and seas guard their germinating civilization, and force its development by the warm interactive life of a contracted territory. Savagery and barbarism mean a scant population, weak tenure of the land, and a fragile bulwark against aggression. Therefore the westward advance of the young republic met only slight hindrance from the indigenous tribes of the country.

The result of this savage occupancy was a country with all the pristine richness of untouched resources. The barrier of the Atlantic was a basis of natural selection among the early colonists, so that in general only the fittest, the robust and enterprising, reached American shores. The abundance of opportunity in this virgin land constantly attracted foreign immigration to reinforce the American population. Easy conditions of earning a livelihood induced early marriages and the consequent large families which became a factor in the expansion of the nation. When with growth of population competition grew stronger and land scarcer, with a fine impatience of these altered economic conditions the dweller of the frontier moved westward to take up unexhausted lands, an unused "range," and also to seek that unconstrained life of the backwoods which has a powerful charm to the natural man. Thus fullness of opportunity bred the migratory instinct, and geographical conditions favored it.

Just as the continental build of North America was unpropitious to the development of small detached nations, so it afforded the natural environment for a few great ones. Its topography is marked by large simple features, — two parallel mountain systems, their two ocean slopes, and between a vast trough-like valley reaching from the Arctic Ocean to the Gulf of Mexico. Small physical features like detached mountain ranges, islands, peninsulas, and deep embayments are scarce. Nowhere is there a highly diversified surface or a highly articulated coast except in the half-submerged lowlands of the sub-polar regions : shore-line and interior are equally characterized by simplicity and unity of structure. This unity is reflected in the few political divisions of the continent.

Eurasia, the land-mass formed by Europe and Asia, whose wide margin is devoted to fringing peninsulas, islands, large isolated mountain areas, and therefore to diversified nations and separate political dominions, has in its interior an unbroken stretch of lowlands and plains shelving down from the continental highlands and extending from the gently swelling uplands of Siberia to the marsh-covered plains of the Baltic. This is the domain of the one continental power of Eurasia, Russia. Expanding eastward from the nucleus of inland empire on the Dnieper and Volga along the line of least resistance through the sparse population of Siberia, Muscovite dominion reached the Pacific by 1645. Westward its growth met likewise no topographical obstruction, but the older population of the Baltic shores, reinforced from Germany and Sweden, retarded its advance till in 1720 Peter the Great made Russia an interoceanic power.

Australia, in consequence of its simple structure, island character, relatively small size, and still smaller habitable area, was destined for a single political territory. The size of North America, some three times as large as Australia, precluded it from becoming in its youth the seat of one empire; but as the general

law of increasing territorial dominion[2] shall continue
to operate, unless counteracted by some superior force,
we may look to see the political union of that which
is geographically one.

All the political divisions of North America reach
from ocean to ocean, with the exception of little San
Salvador. In Mexico and Central America the diminishing width of the continent made such expansion
easy, especially since it was guided by military conquest,
despite the isolation of fending mountain barrier. The
great breadth of the northern section was traversed
in Canada before it was in the United States. Here
physiographic conditions were all conducive to early
expansion, especially when the method was that of the
far-ranging voyageur. No mountain barrier like the
Appalachian system discouraged early progress. The
St. Lawrence owed its great significance in early days to
the fact that it alone opened a way from the east into
the central valley of the continent. Beyond, the Great
Lakes and the low plains of the Saskatchewan carried
the early explorer three fourths of the distance across
the continent to the foot of the Rockies, which bend
rapidly towards the west after entering Canadian territory. Lower passes in the mountains and the absence of
deserts, due to the lower level and narrower width of the
highlands, further expedited Canadian overland advance.
Hence from the Gulf of St. Lawrence to Puget Sound
was the easiest and earliest line of expansion from the
Atlantic to the Pacific. Furthermore, the severity of
the Canadian climate, the consequent smaller incentive
to slow agricultural conquest of the country, and the

greater temptation to the nomadic exploitation of the
fur trade, stimulated the advance towards the west.
These conditions of topography and climate explain the
fact that the Lewis and Clark expedition made their
explorations in the light of data furnished chiefly by
Hudson Bay Company factors.

In the United States geographical features were not
quite so favorable, and yet were not adverse to westward expansion. On the threshold of the continent
stood the Appalachian barrier, but there were five great
routes across or around this, — by the St. Lawrence
and the Great Lakes, by the Hudson and the Mohawk
valley, by the line of the Potomac and the Ohio, by the
Appalachian Valley and Cumberland Gap, and by
the open way through northern Georgia. Beyond, the
mighty Mississippi system, built for intercourse, gave
ready access to the threshold of the Rocky Mountain
portals. The great width of the Cordilleras, flanked
on the east by arid plains which long kept at arm's
length the frontier of continuous settlement, and on
the west by wide stretches of desert, drained by cañon-
cutting rivers unfit for navigation, and scantily supplied
with grass for fodder or wood for camp-fires, greatly
increased the difficulty of American expansion.

But difficulty is always a relative term and must be
measured by the means of overcoming it. The Alleghenies to a race in its infancy were quite as serious an
obstacle as the Rockies to the same race in its prime.
Frontier conditions along the Mississippi and Missouri
trained a people able to cope with the conditions of the
transmontane journey of two thousand miles. Life in

the plains and backwoods had become second nature
to men from whom the need of luxury had been eliminated.   Mere space, unconstrained existence, a buffalo
hunt or an Indian fray was pleasure enough.   In the
large, fresh environment of the American continent
the English race had been born again and now was
animated with the irrepressible vigor of a youthful
people.   A constant change of environment had given
them the adaptability of youth, vast opportunity had
bred the spirit of venture and enterprise.   Nothing
seemed impossible and therefore little was impossible.

The advance from the Alleghenies to the Mississippi
and from the Mississippi to the Pacific was natural.
And in this advance the large and simple structure of
the continent kept the Americans one people.   Each
mountain barrier in turn was regarded as a possible line
of political division until the intercourse made possible
by gaps and approaching streams was developed by the
mechanical power of the people.   The spirit of separation in the young trans-Allegheny commonwealths was
short-lived.   Later on the Pacific coast, the most truly
isolated portion of United States territory, the idea was
current that California and Oregon were destined by
geographic conditions to form an independent state.
This sentiment was voiced in Congress more than once.[3]
And the prompt survey for a transcontinental railroad
after the acquisition of California pointed to the government's fear of losing this peripheral possession.   The
work of construction was long postponed, however, until
in 1862 rumors that the people of the Pacific slope,
tired of waiting for overland communication, proposed

erecting an independent republic, and the report of a
Confederate invasion of New Mexico induced Congress
to lend government aid to the Union Pacific Railroad.[4]

Every forward step in American expansion meant
a more scientific boundary. Behind every process of
rounding out the contour of our political territory lay
the geographical motive. The Mississippi was only a
line of demarcation. Long before the Louisiana Purchase or the discovery of the Columbia, Jefferson saw
the value of a Pacific front. The line of the Rocky
Mountain watershed, especially as reinforced by the
desert beyond, was the highly scientific boundary secured with the Louisiana country; but the ocean is the
only absolute boundary, and beyond lay the Pacific. A
seacoast is valuable chiefly on account of its ports, and
of these the slightly indented shore of the Pacific in
all its vast length south of Puget Sound afforded few.
Entrenched at first on the short strip of coast between
42° north latitude and the Columbia, the United States
had only the mouth of the river for a harbor; but the
vast inlet of Puget Sound was secured by the compromise of the "Oregon Question," San Francisco Bay
by the conquest of California, while by the treaty made
with Mexico in 1848 the international boundary line
was made to bend slightly south of west between the
mouth of the Gila River and the coast in order to
include in United States territory the excellent harbor
of San Diego. The United States was also guaranteed
the free navigation of the Gulf of California and the
Colorado River to the confluence of the Gila.

The Gila River boundary, as fixed by the Mexican

cession, was corrected by the Gadsden Purchase in 1853 of land to the southern watershed. This again represented an advance from an unscientific to a scientific frontier. The Gila River occupies a depression which divides the North American from the Mexican Cordilleras, and which is "one of the most important topographical features in North America."[5] It possesses a strategic value as a natural passway through the mountains and deserts to the west, and therefore its control could not be shared with another power. The easiest trail, as we have seen, approached it by the southern affluents of the Gila. It was the natural route to the Pacific from the coast of Texas, where the highway of the Gulf landed a west-bound traveler a hundred miles farther on his way than a Missouri steamboat which put him ashore at Kansas City. In the Gila depression the first survey for a transcontinental railroad was made as the most obvious route, and here finally was constructed the Southern Pacific line, on the southern side of the valley. The Gadsden Purchase has been decried as a wasteful expenditure of ten million dollars for forty-five thousand acres of poor land almost unfit for occupation. But never was money better spent. The value of the tract is not to be estimated by its fitness for agriculture, but by its strategic importance as a passway to the Pacific which is always open on account of its lower levels and more southern location, when the northern routes are blocked by snow.

The southern boundary is continued by land along the Rio Grande, which serves as a convenient though rather unreliable line of demarcation. From the mouth of this river the United States territory reaches the sea again. The great westward-bending inlet of the Gulf of Mexico creates a southern coast for the United States, which therefore from Key West to the Rio Grande faces South America and not Europe, though washed by Atlantic waters. Confined at first to the strip between the Sabine River and Lake Pontchartrain, the republic with persistent effort worked to broaden her base on the Gulf in order to control the natural outlets of her southern and western country, and give it the absolute boundary of the sea.

The eastern and western boundaries of the United States are alike in being entirely oceanic except for a bit of Maine. The southern boundary is land in its western half, water in its eastern ; so too the northern boundary. The international frontier of Canada and the United States extends from Lake Superior over-land westward along the forty-ninth parallel to Georgia Strait, whence it continues south and west through Haro and San Juan de Fuca Strait to the Pacific ; east-ward from northern Minnesota to northern New York the boundary is defined by the Great Lakes, three short connecting rivers, and the broad channel of the St. Lawrence.[6] Though these " Mediterranean Seas " are fresh water, they have played as great a part as a marine highway in opening up the interior of the country as has the Gulf of Mexico, and that in spite of the fact that they receive no great affluent like the Mississippi to enhance their importance.

Everywhere the contour of the United States terri-tory is simple and as far as possible natural. Any line

separating it from Canada in the west must be artificial
because the valleys, plains, and mountains of the one
country continue into the other without topographical
division; but the forty-ninth parallel has the advantage
of brevity. Moreover, it was easy to draw such an
artificial line of political division in the remote, almost
uninhabited regions of the Northwest. The international boundary of New England, however, is long,
irregular, and follows natural features only for a short
distance along the St. Lawrence watershed, the St. John
and the St. Croix rivers. Contrasted with the northwest frontier and the Mexican boundary, " it marks the
contrast in effect of contact with an old and organized
state and a politically new and unorganized one." [7]

Such were the limits which the republic wrought
out for itself by 1853. The territory which it acquired
was confined almost wholly to the mainland. This is
its distinguishing characteristic. Its islands are few,
small, and close inshore. Island acquisitions were
renounced in the struggle for continental aggrandizement, though some lively diplomatic battles were fought
for the possession of the outlying fragments in Passamaquoddy Bay in the negotiations fixing the eastern
boundary of Maine.

Another geographic factor in the growth of the
United States to a continental power was to be found
in its isolation from Europe. The wide barrier of the
Atlantic which furthered political independence stimulated economic independence and a rapid development
of natural resources more extensive than intensive.
This involved expansion. The young republic of 1776

which turned its back upon the Old World set its face systematically towards its own vast hinterland; it was animated by a continental policy in 1783 in claiming territory beyond the Alleghenies to the Mississippi. This policy was shadowed forth yet more distinctly in the Louisiana Purchase and the Lewis and Clark expedition, and twenty years later reached a novel and final expression in the Monroe Doctrine.

This national principle was enunciated because geographic conditions made it possible. Geographic isolation suggested and dictated its oldest and fundamental proposition, that of our non-interference in the internal concerns of European powers. "Our detached and distant situation," said Washington, "invites and enables us to pursue a different course" from that of Europe with its international entanglements. "Why forego the advantages of so peculiar a position?"[8] The obvious corollary of this proposition was the non-interference of Europe in American affairs,[9] which could best be secured by calling a halt to European colonization on the American continents and the reassertion of European dominion over successfully rebellious colonies. This principle, foreshadowed again by Jefferson, was formally stated first to meet Russian aggressions on the northwest coast, which the United States hoped to make its own through the Oregon claim, before it was embodied in Monroe's message to Congress in December, 1823; so that it was designed to protect the territorial plans of the nation. It was formally adopted as a national policy to protect the Central and South American republics which had been

enabled by a remote peripheral position and the dividing expanse of the Atlantic to secure their independence.   The third proposition of the Monroe Doctrine,
the purpose of the United States to interpose in defense of the independence of any American nation
against any attempted conquest by a European state,
was undoubtedly reinforced by the knowledge that
geographic isolation would enable the champion to inter-
fere successfully.   A broad military base near at hand
on the Atlantic, Gulf, and Pacific, as opposed to the
remoteness of the European powers, even with their
narrow footholds in the Caribbean Sea, would give to
the United States the advantage in any intercontinental
war.

The leadership in the American continents assumed
by the United States in the enunciation of the Monroe
Doctrine has its final basis in geographical conditions.
Identical as to the dominant race, Canada and the
United States differ only in the seats of their respective
dominions; hence climate, soil, location, and natural
features have been the differentiating factors in their
history — the determinants.   In the contest of expansion which resulted in this leadership, Great Britain
was the most dangerous competitor of the United
States; it was an American power with large interests
in this continent which could be seriously marred by
new appropriations of territory by its republican rival.
But the center of British power was extra-continental
in relation to America : the broad interoceanic belt of
Canada was not enough to cinch the western continent.   " The United States had all its forces on the

ground and no distracting interests in other parts of the world. Hence in all the debatable territory England gave way. Texas was annexed without opposition; Oregon was divided; California was conquered." [10]

The isolation of North America gradually eliminated the lesser European rivals of the United States in the Western Hemisphere, and enabled her by securing some of their previous holdings on the mainland to carry out her continental policy of compact growth, which made for strength in the struggle with England for continental control. The compact political territory, having little exposed surface as frontier, is less vulnerable than the widely scattered political area, and can expend all its powers at will in any one direction, just as Russia throws its great weight to-day against the wall of the Hindu Kush, to-morrow against the buffer of Manchuria. The United States has in general directed its efforts at only one thing at a time and that with success, because its energies have not been dissipated as have England's.

Geographical location has been the most powerful factor in this leadership; it has given the United States advantages over both Brazil and Canada, the only other two large territorial powers of the Western Hemisphere. Canada has, like the American republic, an interoceanic position, but it is too far from the center of things to make its influence felt widely, and has too severe a climate to permit the development of the dense population necessary for strength. Brazil has the fertility and area, but its tropical location will always limit the energy of its people, even if we leave out of consideration the limitations for leadership native to the Latin races; and Brazil's want of a Pacific and Caribbean seacoast must confine its sphere of influence largely to the Atlantic.

The location of the United States makes it the master of the situation. It alone has a location wholly in the temperate zone stretching from ocean to ocean. Thus it has a population energetic by reason of an invigorating climate and multitudinous because of the large and fertile area of the country. Its extension to the Pacific brings it into relation with the western states of South America, while its long coast-line on the Gulf of Mexico makes it the leading Caribbean power. Its position in the North Atlantic opposite Europe makes it the great maritime doorkeeper of the western world. Thus this central belt of the North American continent was geographically determined as the seat of control of the new hemisphere. This destiny was fixed by its colonization by the sturdy Anglo-Saxon race.

The evolution of the place of the United States in world politics is interesting. In its colonial days, America, as merely the western periphery of Europe and an appanage of the Old World, was subjected to all the international forces which swayed her sovereigns. European wars had episodic campaigns on American soil, and provisions as to American territory figured in all their treaties from Ryswick to Paris (1763).[11] For a short time after the Revolution the young republic was embarrassed by some responsibilities to France, heritages from the time she formed a part of the European body politic, which led France to think she might be used as an ally against England. But in the isolation of her new environment, the United States quickly threw off her European connections, worked out her destiny as the foremost American state, became an interoceanic continental power, then a hemisphere power, dominating the international affairs of North and South America.

For a long time this domain satisfied her. Geographic isolation was so strong in its suggestion that there was no temptation to extra-continental political activity; but even in this period of apparent quietude the seeds of further expansion were germinating,— expansion which was to lead the United States beyond the confines of her hemisphere, now grown too narrow, to take her place as a world power. Isolation gave to the infant giant that protected position which fostered political development on a large scale, just as the segregating environment of the Greek city-states achieved for those a similar development on a small scale, till in these latter days, issuing from the seclusion of its hemisphere, the United States has participated as a world power in the affairs of the Pacific and of Asia.

The size of the United States has been determined largely by the continental build of North America. We have seen how its simple structure facilitated western expansion and preserved the unity of the people. We have seen how the ranks of the future nation were drawn up in compact form along the narrow Atlantic littoral between the mountains and the sea, how the line of march was flanked by the Great Lakes on the north and the Gulf on the south, how the Mississippi carried the outbound conquerors southward to the Gulf and the Ohio tributaries carried them north to the Lakes; how the length of the Mississippi became the base for the next advance. The imperfect geographical knowledge of 1783, supposing the sources of this river to be northwest of the Lake of the Woods, by a happy mistake fixed the incipient point of the northwest boundary beyond the head waters of the Mississippi, and thereby increased the dimension of the United States along the ninety-fifth meridian.

Confined between the Lakes and the Gulf, the natural outlet of growing powers was towards the west. Had the continent been wider, its width would have been mastered, though perhaps more slowly; had the Mississippi been longer, its greater length could have been as easily controlled, providing its geographical nature remained the same. The effect of the continental build, combined with the constricting influence of the Great Lakes and the Gulf, has been to give the United States a predominant east and west direction, with a dimension twice as great as that from north to south; to keep it within relatively narrow zonal limits, away from tropical heat or polar cold, and yet to lend it the enormous area of three million square miles.

Geographic conditions further conspire to make by far the larger part of this area available. Through the great central trough of the continent sweep the cold winds of the north and the warm moisture-laden breezes of the Gulf, modifying in turn the climate of Minnesota and of Texas. The narrow width and lower level of the northern Rockies admit the Pacific winds which bring warmth and moisture to Montana and the Dakotas. There is probably no other continuous political area of like size which contains so large a proportion of territory adapted to the habitation of man. The size of a land is to be measured finally in terms of the population it can support and bring to a high degree of civilization ; amount and effectiveness of population is the measure of power.

The effect of the size of their country can be traced in the ideas of the American people, which are marked by a certain largeness and daring. The small territorial standards of the early European settlers here became profoundly modified by American continental conditions. The mere area of the individual states increases from the east towards the west. The commonwealths of New England seem pigmies in size compared with the trans-Mississippi states. There are twenty-six states in the smaller half of the country east of the Mississippi, and only twenty-three states and territories west of it. The greater aridity of the western states, their consequent sparsity of population, and also the rapidity with which immigrants spread over them account in part for this larger area. The Senate will always be the legislative stronghold of the eastern states, where representation is based upon units, not on population. The West can therefore look for growing power only in the House, where increasing population in the young states will increase their representation. In the recent discussion of the admission into the Union of the remaining western territories, the proposition was made by eastern statesmen to unite New Mexico and Arizona in order to limit their representation in the Senate and avoid the evil of more "rotten boroughs" like Nevada, destined by nature for a scant population ; and to admit Oklahoma on condition that it shall one day incorporate in itself the Indian Territory, when under the new Indian policy that shall cease to be a tribal reserve.

The large area of this country, the mere space to be overcome whereby time might be saved, has brought the United States to the front in improved methods of transportations, whether in railroads, river steamers, ocean liners, or in rapid battle-ships like the Oregon. The numerous large rivers which we have had to span have made us the great bridge-builders of the world. Hence the successful competition of America for the Atbara bridge in Africa. All other American industrial enterprises manifest the same magnificence of scale, especially agriculture in the abundant lands of the West. On one wheat farm in Dakota a man ploughs a straight four-mile furrow; down its length takes from breakfast to his noonday lunch, and back to the start brings him home just in time for supper. In Texas a cattle ranch often embraces a domain from one to two hundred thousand acres enclosed by a single wire fence.

Everywhere one notices a certain largeness of view in the ordinary Westerner. Even when uncultured and crude from lack of opportunity, he never takes a contracted view of things. He measures things with a big yardstick. The nomadic instinct is still in him, handed

down by his emigrant forbears. Wherever he is found
he has always come there from somewhere else. Hence
he is never provincial, and he is intensely, broadly
American. Distance never appalls him. If he is a
Californian he " steps over the mountains into Nevada,'
literally on foot if the fancy take him, with as little ado
as if he were walking around the corner; or he crosses
half the continent by rail for a two days' visit at
Chicago. His point of view is therefore bred of his
geographically wide experiences and his intercourse with
the other highly mingled populations of the western
states. The areas of provincialism in the United States
are few, and none of them are to be found west of the
Mississippi River. The danger in the American point
of view lies in the tendency to confuse bigness and
greatness.

" The struggle for existence is a struggle for space,"
says Ratzel. Abundant space in the United States has
meant abundant opportunity and a chance for all to
rise; it has developed in the Americans a powerful initiative and encouraged the democratic spirit. Thus as the
isolation of North America helped the early colonists
divest themselves of European monarchical principles
of government, so the size of our country has kept
classes and masses on a nearly equal footing by equality
of opportunity. Everywhere the " four hundred " tends
to recruit itself from the four million near by.

The transcontinental expansion of the American
people has made them masters in " the struggle for
space." The question is not remote, How will they
use their power so acquired in the future ?

NOTES TO CHAPTER XII

1. Shaler, Nature and Man in America, pp. 168–172.
2. Ratzel, Politische Geographie, Dritter Abschnitt.
3. McMaster, History of the People of the United States, vol. v. p. 26. Bancroft, History of Oregon, vol. i. pp. 350–380.
4. E. E. Sparks, Expansion of the American People, p. 368.
5. Ratzel, Politische Geographie der Vereinigten Staaten, p. 33. 1893.
6. Ibid. p. 35.
7. Ibid. p. 36.
8. Washington's Farewell Speech, September, 1796.
9. John W. Foster, A Century of American Diplomacy, p. 441. 1901.
10. A. B. Hart, The Foundation of American Foreign Policy, p. 36. 1901.
11. Ibid. p. 3.

CHAPTER XIII

THE GEOGRAPHY OF THE INLAND WATERWAYS

THE simplicity of build which gives to the North American continent its few large features of valley, plain, and mountain range has been the chief determinant in producing its few large groups of navigable rivers; for these are always the product of big, well-watered countries of simple form and gentle slopes. The complex, fragmentary topography of western Europe is too much broken up to develop a great river system. The continental structure of Australia favors a large central system like the Mississippi; but the location of its mountain ranges on the eastern rim, where they intercept the moisture of the trade-winds, robs the wide, interior valley of its water. Africa has thousands of miles of navigable waterway high up on the surface of its plateau, but the steep escarpment of this mesa continent bars every river approach from the sea with falls and rapids a few miles back from the coast.

Everywhere in North America, except in the trans-Rocky region, an easy gradient marks the descent from highland to sea. The natural obstacles to navigation are few and slight. The Ohio from the edge of the Allegheny Mountains to its mouth has only one partial interruption to navigation in the so-called "Falls" at Louisville, over which, however, steamers pass for a great part of the year. The Missouri pours along its smooth, yellow tide for 2700 miles from the Great Falls in western Montana, where it plunges down the first terrace of the Rockies, without other break to its mouth. The gradient of the Mississippi's bed becomes pronounced only above St. Paul, while insignificant watersheds barely suffice to bar the sources of its tributaries from the Laurentine Lakes, or from the Red River of the North which, through Lake Winnipeg, finds its way to Hudson Bay.

Everywhere east of the Rockies the chief drainage systems of the continent approach very near on some low divide at one or several points. Here were located the portages or "carries" which extended the network of inland commerce in the days of bark canoe and beaver-skin cargo. Here later were built the canals which answered to the demand of spreading settlement and increasing trade, and linked the Atlantic rivers with the St. Lawrence, the Great Lakes, and the eastern branches of the Mississippi. Thus the topography of the country enabled even these abundant water connections to be vastly extended by artificial means. The waterway of the Hudson River, Erie Canal, and Great Lakes has an east and west extent of eighteen degrees of longitude, reaching from New York to Duluth. The chief lateral branches of the Mississippi, the Missouri and Ohio, afford river connection from Fort Benton in western Montana to Pittsburg, a distance of 1800 miles in a straight line; and the canal links between the Ohio and Lake Erie extend this water route to the Atlantic. The east and west direction of the Ohio and

Lake route was a strong factor in the union of the early East and West, counterbalancing the centrifugal influence of the Mississippi. And earlier still, the close approach of the Potomac and Ohio, and the ease of intercourse along the highroad between their respective heads of navigation, was one of the chief reasons for locating the national capital at Washington.

The Atlantic rivers, where they are navigable, whether naturally or rendered so by canals, facilitate chiefly an east and west connection. The Mississippi system is most important as a highway between the northern and southern borders of the country. Northward from the line of the Ohio and the Missouri, as far as the great bend at Kansas City, this north and south connection is manifold. That of the main stream is reinforced by the upper Missouri to the west, and to the east by every tributary of its own or of the Ohio having canal communication with the Lakes. Southward from the Ohio only the main stream of the Mississippi forms the conjunction with the Gulf. The second parallel route, known and used by the canoes of the early traders, from the great bend of the Tennessee over the portage south to the Black Warrior or Tombigbee rivers, has not yet been utilized by the construction of a canal; but an outlet from the Tennessee to tidewater in Mobile Bay is a proposed improvement of our inland navigation, conditioned only by the problem of an adequate water supply on the summit level. The value of the Mississippi as a waterway is enhanced greatly by the fact that it traverses the United States across the constricted area between the Lakes and the Gulf, while these two

mediterraneans, fresh and salt, find their value enhanced
in turn by the connecting waterway of the mighty
stream.[1]

If we leave out of consideration the abrupt eastward
projection of the state of Maine, the ninety-seventh
meridian is found to mark the medial line of the United
States and also the western limit of abundant inland
navigation. Corpus Christi, the western terminus of
coastwise navigation on the Gulf, and the heads of navigation on the Red River and the Arkansas are located
almost exactly on this ninety-seventh meridian. Beyond this line only the Missouri affords a navigable
course to the foot of the Rockies. Scanty rainfall and
shallow, shifting beds render the other streams unfit
for transportation.

The western part of the United States embraced in
the broad band of the Cordilleran upheaval, in consequence of a very complex topography combined with
extreme aridity, has developed no long river systems
navigable from the sea. Aridity has produced the prevailing cañon formation and the scanty volume of
water which will float only the shallowest boats. The
lower Colorado, after it issues from the Grand Cañon,
will carry vessels of eighteen-inch draft from Eldorado
in southern Nevada four hundred miles to its mouth.
But this sole river outlet to the south has little value
because of the desert country and scanty population
which it serves. The imprisoned streams and lakes
of the Great Basin have been of service merely in furnishing a line of oases which directed the route of the
California Trail. Only aridity, due to the encircling

mountains, has prevented these streams from developing the volume necessary to cut away the slight oppos-
ing barrier, and uniting to a system to find an outlet
along the general slope of the Basin to the north.

The gentle curve of the Sierra Nevada and Cascade
Mountains marks the eastern limit of the inland waterways of trans-Rocky America. What is found behind
the barrier of these ranges counts for little or nothing,
and the aggregate of that beyond is only small. Puget
Sound, with its 19,000 miles of shore-line, owing to its
predominant north and south direction, finds its most
eastern point only about one hundred and twenty miles
from the sea. The Columbia, accessible to ocean vessels a hundred miles from the Pacific, has continuous
navigation only a hundred miles farther to the Dalles,
where the river rushes down through the narrow gateway of the Cascade Range. Its plateau course farther
inland is navigable for a great many miles within the
state of Washington, but is little available on account
of its inaccessibility from the sea. The long valley
formed by the Sierra Nevada and Cascade Mountains
on the east and the Coast Range on the west is
occupied by the Willamette, navigable for steamer and
river craft for one hundred and twenty-five miles, and
further south by the submerged river mouth forming
San Francisco Bay with its tributary streams of the
Sacramento and San Joaquin, which together afford a
north and south waterway aggregating three hundred
and fifty miles. But like Puget Sound, all these rivers
except the Columbia, owing to their location in a great
longitudinal valley following the axis of the mountains,

reach points less than one hundred and twenty-five miles back from the Pacific and form, as it were, only a coast road at the foot of the vast Cordilleran upland.

The abundance of inland waterways in the East and the paucity in the West has greatly differentiated the economic history of these two sections of the United States. Every country of large extent finds, next to the fertility of its soil, its navigable rivers the most important factors of its early development. The one is fundamental to the production of wealth and the other to its distribution and exchange. Small countries with deeply indented coast-lines, like Greece, Norway, England, and New England, can renounce the advantage of big river systems; but in Russia, Argentina, Venezuela, the United States, Canada, Egypt, India, and China, the history of the country, political and economic, is indissolubly connected with that of its great rivers. Large areas mean great distances, hence long lines of transportation, which are most easily and cheaply furnished by natural waterways; [2] and the complete exploitation of such natural advantages has been in all times a condition and measure of commercial preëminence. The strength of Venice and the Lombard League found its geographic basis in the Adriatic, the roads across the Alpine passes to the north, east, and west, and the navigation of the Po and Adige. The League of the Rhine towns flourished by reason of the Rhone-Rhine highway across western Europe. The Hanseatic cities from Bruges all the way east to the Russian Novgorod owed their commercial supremacy largely to the great rivers behind them from the Scheldt to the Volga. [3] In

France the signal for a marked industrial and commercial development was the construction by Colbert in the seventeenth century of a series of canals across the central plateau of the country, designed to unite the sources of all its radiating streams and perfect a wonderful system of inland waterways. The interior trade of China to-day moves wholly by river and canal; and this country remains the classical example of waterway supremacy. The foreign "spheres of influence" there are indicated generally in terms of the great Chinese river, because these "spheres" are commercial areas tributary to the river highways.

Early settlement and expansion of population in the United States, falling as it did in a period prior to the rapid development of transportation by means of railroads, and being confined to the well-watered Atlantic slope of the continent, developed, as we have seen in the previous chapters, largely by reason of waterway facilities. Line of river and lake invited and directed the westward movement, distributed the outbound settlers along remote tributaries, and the next year conveyed the produce from the newly opened lands back to the seaboard markets. For every outgoing stream of people to the West, there was a return current of trade, steadily increasing in amount, variety, and value, which moved along these same channels. Later, when railroads were introduced, the waterways remained as the cheap competitor of the steel roads, regulating their freight charges from the Lakes to the Gulf and from the Atlantic to the ninety-seventh meridian.

In the trans-Missouri country with its scanty equipment of waterways, geographic conditions have differentiated greatly the history of transportation. On the margin of the arid belt, canoe, flatboat, and steamer were discarded; pack-horse, ox-cart, and mule team treked across the endless stretch of prairie and desert. For the West there was nothing between the creeping pace of the canvas-covered wagon and the railway express, — that too in a country where time was money and life, if an early snow threatened to block the mountain passes, if the scanty pasturage in the desert oases was all consumed, if the food supply had to be reduced to lighten the burden of the exhausted animals, whose dragging, trembling pace made the desert twice as wide, the hope of water twice as remote.

Here was the land where the railroad found no competitor, a land moreover of "long hauls" and "through freight," except for the sacks of wool from the scattered highland ranches, and the cases of canned vegetables dropped off at the lonely shacks of the desert road-keepers. Wherever the cheap highways of rivers and lakes are lacking, railroads must develop so much the more rapidly.[4] The sparsity of population along their route diminishes the local business and hence the profits; but the absence of waterway competition enables them to formulate their policy untrammeled, fix their own tariff, and follow the lines which the interests of commerce dictate. Railroads have assumed an incalculable importance in Australia because of the paucity there of permanently navigable streams. Similarly in South Africa, an extensive area without water communication, Cape Colony, Natal,

Rhodesia, the Orange River Colony, Transvaal, and Portuguese East Africa, have been forced to construct over fifty-five hundred miles of railroad in spite of the youth of the country and the sparsity of its white population. Even this large relative aggregate is inadequate to the existing commercial and military needs, and numerous extensions are already projected.[5]

In a like manner, our West has developed railroads at an enormous rate, in spite of sparsity of population, owing to the economic necessity of connecting two opposite coasts which serve two areas of trans-oceanic commerce. Freed from the regulating control of competing waterways, the western lines are in the position of a monopoly. Hence the greater danger in their "community of interests" system and the bitter fight against it in the western states. The nearest rival of the overland roads is the Straits of Magellan, though a partial competitor is found in the sea and land route of the Gulf of Mexico and the Isthmus of Panama. The first serious rival will be the Isthmian Canal, which will make itself felt as far north as the Canadian Pacific Railroad. Owing to this lack of waterway competition in the far West, surprising importance is acquired by such small rivers as the Sacramento and San Joaquin in California, which are navigable for large steamers only to Sacramento and Stockton respectively, but which are used extensively for passengers as well as freight, even at the cost of personal inconvenience, in protest against the monopoly of the Southern Pacific Railroad.

Twenty-five years after our trans-Rocky expansion

had begun in earnest, an overland railroad united the Pacific with the Atlantic seaboard. Transportation development in the far West sprang from the trail to the steel track, omitting the intervening stage of constructed highroad on the one hand and unretarded by the presence of waterways on the other. In the eastern half of the United States, land routes felt everywhere in the early days the competition of inland navigation. The abundance of excellent waterways tended to discourage the construction of highroads except such as led across mountain barriers, like the old coach-road from Philadelphia to Pittsburg, the Cumberland National Road from the head of canal navigation on the Potomac to Wheeling, and the Wilderness Road across the southern Appalachians; or across the axis of the prevailing streams, like the early post-road from New York and Philadelphia southward along the coast to Charleston, South Carolina, or the extension of the Cumberland National Road from Wheeling westward across the northern tributaries of the Ohio through Columbus, Indianapolis, Terre Haute, and Vandalia. New England, owing to its lack of inland navigation, was the part of the country which earliest developed a complete system of turnpikes and later of railroads; and even to-day it is the favorite haunt of the bicycler, who elsewhere suffers from the fact that the country slighted the highroad phase of communication.

Abundant waterways therefore mean prevailing water transportation in the early stage of a country's economic development; the larger the size of a country and the greater its distances, so much the stronger its need of inland navigation, the greater the development and the longer the survival of the same. Owing to the narrow width of the Atlantic plain, its southeastward slope, and the " fall line " never very far from the coast, the career of our Atlantic rivers as avenues of transportation has been short in period and limited in extent. With the exception of the Hudson, they have served only east and west communication. Intercourse between North and South was delegated to coastwise navigation, which here performed the function of the Mississippi boats in the interior of the land.

The importance of these Atlantic streams was greatest in colonial days, because, on account of the small sea craft then in use, there was less distinction technically between inland and marine navigation. Furthermore, these rivers, being small, have suffered relatively more from that necessary contraction of navigable waterways which has resulted everywhere from the substitution of large steamers for the shallow boats of the early period. Finally, since in all water transportation the cost falls largely to the expense of loading and unloading, the small Atlantic rivers are under the economic disadvantage of the "short haul." Hence only those have maintained their importance which, like the Hudson, St. Lawrence, and lower Delaware are navigable to seagoing vessels.[6]

It was in the great valley of the Mississippi that inland waterways proved to be a most powerful economic and political factor. This basin, embracing 1,240,000 square miles, containing 15,410 miles of streams navigable for present large river craft and having therefore a much greater mileage in the day of canoe and pirogue, early exploited its advantages of water communication, all the more because of the rapid expansion of settlement throughout the valley, the increasing distances to be covered, and the sparsity of its population, which could not bear the expense or even furnish the labor for road-making. In the pioneer days it was downstream travel for the emigrant settler to his new home along the Ohio, the Tennessee, and the middle Mississippi, as later for his produce to the market at New Orleans; but increasing economic development with its more extensive exchanges made an insistent demand for more effective upstream navigation, as for more rapid water communication in general. Hence the introduction of steam as a propelling power solved the problem of the swift western waters.

In 1811 the banks of the Ohio first echoed to the stroke of a steamboat's paddle-wheel; and between 1812 and 1836 eight trans-Allegheny states were admitted to the Union. Shipbuilding became an important industry, centered chiefly in Pittsburg, Cincinnati, Louisville, and St. Louis, though carried on also at Wheeling and Marietta. All of these cities were situated on the east and west waterway of navigation and traffic; they commanded abundant lumber supplies and especially the iron necessary for machinery and construction which came from the mines on the upper Ohio system; and finally they commanded skilled labor from the influx of northern settlers with industrial habits such as were not to be found in the slaveholding region farther south. Industrially the line of the Ohio was a boundary zone, embracing in its lower course the economic systems of the North and South. New Orleans essayed steamboat building, but never developed it as did the points along the Ohio. Her geographical location, commanding a sea-route to the manufacturing centers of Europe and the North Atlantic states, and a great river highway for the downstream transportation of her food supplies, made her independent of all craft but sailboat and barge. The upstream men needed the steamers.

In the year 1818 five steamboats were built at Pittsburg, one at Wheeling, four at Cincinnati, and four at Louisville, or fourteen in all; the next year Louisville built twelve, Cincinnati six, Pittsburg two, Wheeling one, and New Orleans two, in all twenty-three for the year 1819. The stimulating effect of the long swift currents of the western rivers upon the building of steamboats may be seen by a comparison of the statistics of this industry in the Mississippi valley and along the Atlantic coast and the Great Lakes, where sailing vessels were adequate to the demand of that time. Up to 1820 there had been built upon the Great Lakes four steamers which measured altogether 831.84 tons, and on the Atlantic coast (exclusive of New England) fifty-two steamers measuring 10,564 tons, as against seventy-one steamers measuring 14,207 tons on the rivers of the Mississippi valley. Even in the next decade, when a continuous band of relatively dense population reached to the southern point of Lake Huron, and the Erie Canal had begun to stimulate traffic on these inland seas, only eight steamers were built on the Lakes.[7]

In 1834 there were on the western rivers two hundred and thirty steamboats, with an aggregate tonnage of 39,000, and in 1842 there were four hundred and fifty boats measuring 90,000 tons. The number had reached six hundred in 1843, and twelve hundred in 1848. Such a remarkable development reflects the importance of great waterways in a new, vast, and highly productive territory. The vessels were built for the least draught of water, and to stem the strong currents of the rivers. They carried heavy freight, which included both agricultural and manufactured products. On the large, elegantly equipped passenger boats, which made regular winter trips between Cincinnati or St. Louis and New Orleans, gay social life with nightly balls gave to river travel on the Mississippi a local color such as it acquired nowhere else. Less than fifty years from the ripple of the fur-weighted pirogue, paddled along by the voyageur with his buckskin clothes and half-savage habits, to the monotonous splash of the big paddle-wheel and the floating palace with its lights, music, and the polished society of the generous Southland.

The business of forwarding the river commerce absorbed much of the capital and the best energy and talent of the larger Ohio river cities. This was especially the case with Louisville, which exploited the advantages of its location at the natural obstruction made by the falls to continuous navigation. Low water on the falls stopped the passage of vessels over these rapids. Louisville therefore was the natural port for the upper Ohio, and the head of navigation for the

lower stream. Finally the increasing importance and
value of the river traffic rendered it advisable to open
a canal around the falls in 1830; but the tolls charged
were so exorbitant that boats plying between Louisville
and the South loaded and unloaded at Portland just
below the rapids, and the peculiarities of Louisville's
geographical location remained unaltered. Meantime,
its steamboat captains, who came from the best elements
of the community, were amassing fortunes.

Similarly St. Louis, by its dominating geographical
position at the plexus of the western waterways, became
an important agent in the commerce of the Mississippi.
Below St. Louis the depth of the Mississippi is six feet
or more, above it is only between three and five feet.
This fact has differentiated transportation on the upper
and lower river and made St. Louis a point of reship-
ment. It became the outfitting-point for the few early
steamers which, between 1813 and 1844, carried supplies far up the Mississippi to the Indian traders operating in its upper courses and the troops stationed at
Fort Snelling, near the Falls of St. Anthony; for the
boats running on the Missouri River to Independence
and Council Bluffs with supplies for the overland trails;
and for the steamers of the American Fur Company,
which up till 1850 were the only ones to penetrate the
wilderness of the upper Missouri above Council Bluffs.
Once a year they set out to gather in the harvest of
the winter's hunting and trapping, advancing farther
and farther upstream as the fur-fields receded, — in
1831 to St. Pierre in South Dakota (1330 miles), and
in 1832 to Fort Union just above the mouth of the

Yellowstone (2000 miles), which long remained the head of navigation. As population began to spread up the line of the Missouri in western Iowa between 1859 and 1860, a weekly boat ran from St. Louis to Sioux City, the head of navigation for larger vessels; and when this place became the terminus of the Sioux City and Pacific Railroad in 1868, a steamboat company operated in conjunction with it, carrying private, military, and Indian freight to Fort Benton (2663 miles), the final head of navigation.[9]

Meanwhile the lines of St. Louis commerce were increasing in number and extent, and the amount of its traffic growing from day to day. In 1845, when river transportation was at its height and had not yet begun to feel the competition of railroads, a report on the river trade of St. Louis during that year records 2050 steamboat arrivals, aggregating 358,045 tons, besides 346 keel and flat boats. Of the steamers, 250 came from New Orleans, bringing fine merchandise of foreign or New England manufacture to exchange for the flour and bacon of the more northerly states; 406 came from ports along the Ohio and its tributaries, laden with cargoes of agricultural products for the St. Louis market, or with manufactured goods which had come in from the Atlantic seaboard by the eastern canal routes; 298 came from ports along the Illinois River, 643 from the upper Mississippi, 249 from the Missouri River, and 204 from other ports, chiefly Cairo and intermediate points.[10] These figures reflect the commercial development attained by a young country because of its abundant waterways. The same story comes from another source. Cairo, at the confluence of the Ohio and Mississippi, had an excellent location for keeping tally on the river traffic. A record of all vessels passing that point during the year 1840 stated the number as 4566.

Following the progress of steam navigation and the rapid settling of the country, the growth of commerce on the Mississippi system was enormous, and early initiated a demand for national improvements on the watercourse. These were inaugurated by surveys as early as 1819, but the actual improvements began in 1827 and followed rapidly for the next decade. The beneficiaries at this period were the Missouri, Ohio, Wabash, Cumberland, Tennessee, Arkansas, Red, Bayou Teche, and the main Mississippi with the passes at its mouth. In the early days the depth of water in the outlets sufficed for existing vessels, but in 1875 operations were begun to make New Orleans accessible to modern ocean steamers. A seaport at a great river's mouth is always the converging point for many lines of inland and of marine navigation. The interests of commerce demand that the contact here of river and ocean should be as extensive and perfect as possible. Hence the outlet of a great river system presents a natural field for the expenditure of large sums to this end, especially in a delta region where the branching of channels multiplies the avenues to the sea, but, owing to their tendency to silt up or shift, require artificial aid to keep them open. Therefore in the delta regions of the Rhine, Scheldt, Po, and Adige, as earlier of the Nile, Ganges, and Chinese rivers, water communication with the sea has been extended and secured by means of canals.[1] Similarly in the United States, the government has expended large sums for the improvement of the passes and bayous of the lower Mississippi. Appropriations have secured the navigability of the Barataria, Atchafalaya, Terrebonne, Black, and Lafourche bayous, while canals give New Orleans additional outlets to the Gulf through Lakes Salvador, Ponchartrain, and Borgne. In its network of waterways the Crescent City resembles in some degree Bruges, Amsterdam, Arles, Cairo, and other delta emporiums.

As the dividing channels of the lower stream suggest this extension of navigable outlets to increase the lines of communication with the great ocean highway, so the spreading branches of a river source which approach or meet other such head waters over a low divide suggest the extension of inland navigation by the union of two drainage systems through canals.

The gentle slopes which so generally characterize the eastern half of the United States, and the long, low line of the watershed separating the drainage basin of the Great Lakes from that of the Hudson and of the Mississippi, made feasible a system of canals completing the "Great Belt" of navigation from St. Lawrence and New York bays to the Gulf. Every early canoe route of river and portage over lake-dotted or swamp-covered watershed to other river or inland lake beyond pointed the line of a possible waterway, whose first surveyor was the redman coming to sell his winter's hunt of fur, or the voyageur outbound with gaudy merchandise to trade in the Indian camps of the interior. Every such trade-route from the Hudson and the Susquehanna

to the Northern Lakes, and from the long chain of lakes
to the springs of the Ohio and Mississippi, has been the
line of a projected or accomplished canal. And just
as in the days of the fur trade, the passway of the
Mohawk, Wood Creek, Lakes Oneida and Ontario enabled the Dutch and English to draw the peltries of
the great Northwest away from the St. Lawrence down
to the Hudson, so the Erie Canal has enriched New
York with the wheat, lumber, and animal products of the
Lake region to the detriment of Montreal and Quebec;
and so Canada to-day is projecting a short-cut ship
canal from Georgian Bay through Lake Nipissing to
the railroad down the Ottawa River, along the line of
the voyageur's interior route in the old days, in order
to secure its hold upon the growing wheat export of
Manitoba and the Northwest Territories. [See map of
Great Lakes portages.]

Men like Washington and Jefferson, who were alive
to the political as well as the commercial benefits of
communication across the country, saw the possibilities
of canals in completing the circuit of vast interior waterways.[12] "Extend the inland navigation of the eastern
waters — communicate them as near as possible with
those which run westward; — open these to the Ohio;
— open also such as extend from the Ohio towards
Lake Erie, — and we shall not only draw the produce
of the western settlers, but the peltry and fur trade of
the Lakes also to our ports . . . binding the people to
us by a chain which can never be broken."[13] In 1786
Congress declared all the portages between the Ohio
basin and the Great Lakes common highways. Albert

Gallatin in his famous report of 1808 [14] pointed out the topographical features of the three low watersheds east of the Mississippi rendering canal communication between the East and West possible, and also the four necks of land at the base of the Atlantic peninsulas which could be pierced to give a shortened and protected passage for coastwise navigation from Boston to South Carolina. Many of the 'canals proposed in his system had already been attempted or partially constructed by private corporations. The Dismal Swamp Canal (1785 to 1794) was even then bringing shingles from the forests of North Carolina in barges of two feet draft to Chesapeake Bay, and the Carondelet Canal (1808) had just given New Orleans an outlet to Lake Ponchartrain. When in July, 1825, the Delaware and Hudson Canal from Port Jervis to Kingston was begun, the Delaware and Chesapeake was well under way. Raritan Bay was linked to the Delaware River at Bordentown by an artificial channel across the neck of the New Jersey peninsula in 1838.

The burning question for the consolidation of the country was communication between the East and the West. The consideration of Gallatin's plan for a waterway around the southern end of the Appalachian barrier by a canal five hundred and fifty miles long, from tidewater in Georgia across the upper courses of the Chattahoochee and Mobile rivers to the Mississippi, was postponed by the Indian occupation of this country till the introduction of railways had ushered in the final stage of transportation development. The direct course of the Potomac from the heart of the Allegheny

Mountains suggested a canal over the watershed to the sources of the Ohio. The channel was constructed from Washington to Cumberland, but there abandoned as impracticable. The zigzag course of the Pennsylvania rivers across the Appalachian system, the general parallelism of their flow from northwest to southeast, and the interlocking of their tributaries in the high mountain valleys between suggested the utilization of even this mountain topography for canal communication with the Ohio. The Schuylkill, Lehigh, and Lackawanna were canalled along their lower course, but no means were found to extend any one of these links westward across the intervening range to the Susquehanna; so for this stretch a railroad between Philadelphia and Columbia was substituted. The transmontane canal began therefore at Harrisburg, a little farther up the Susquehanna, and followed this stream and its western tributary, the Juniata, to the base of the Allegheny range. There another short railroad over the summit (2491 feet elevation) made the connection with the western canal down the Conemaugh and Allegheny rivers to Pittsburg. This Portage Railroad, as it was called, took the place of a proposed tunnel canal forty-two miles long.

This circuitous and complex route, made up of river, canal, and railway, though valuable as an outlet for the mineral products of Pennsylvania, could not serve the purpose of a trans-Appalachian canal. That destiny was left for the Mohawk depression, where the mountains drop to an elevation of only 445 feet above ocean level and where the inland objective was not a single river system like the Ohio with a navigable depth of five feet, but a chain of interior seas with coast-line of nearly 4000 miles, navigable to large vessels propelled by steam or sail, and affording in their turn canal communication to all the vast territory contained in the mighty angle of the Ohio and upper Mississippi.

The first suggestion of the Erie Canal came as a proposition in the New York legislature in 1785 for the improvement of inland navigation at various points between Albany and Oswego. As the idea of a continuous canal developed, Lake Ontario was still regarded as its western terminus. This was the canal embodied in Gallatin's report, though even in 1808 the superiority of a longer route direct to Lake Erie was discussed and a proposition for its survey made in the New York assembly. The exploration of this route disclosed the fact that the series of elevated lakes in central New York would exclude the necessity of constructing expensive reservoirs on the summit level, since the canal would occupy the depression which received the outlet streams of these natural basins. Here was everything needful for a cross-country waterway of 363 miles; a deep river leading northward from one of the finest ports in the land, a tributary stream from the west flowing down from a broad low mountain gap, lake feeders on the summit, and a terminus on a line of inland seas.

The competition of Lake Ontario and the St. Lawrence was already deflecting the greater part of New York's internal commerce to Canadian ports, and the Susquehanna was carrying off much of its produce to Chesapeake Bay. Then came the War of 1812, demonstrating the strategic necessity of a protected, wholly American line of water communication with our western frontier, one which should avoid the exposed stretch of Lake Ontario and the obstacle of Niagara Falls. Military as well as commercial considerations dictated the wisdom of extending an arm of this canal up to Lake Champlain, which in the recent hostilities had occupied as dangerous and isolated a position as had Lake Erie at the time of Perry's operations, and which possessed a further strategic importance as a natural avenue from the Canadian boundary. Another branch from the main canal at Salina to Oswego on Lake Ontario completed the hold of New York on its frontier lakes. Thus the Hudson River became the channel through which three distinct streams of commerce moved down to the port in New York Bay, when 1825 saw the formal opening of this great interior waterway. The next year, nineteen thousand boats and rafts from the Erie and Champlain canals filed past West Troy on their way down the river. Shipbuilding grew up as a regular industry on Lake Champlain, to furnish transportation for the staves, shingles, boards, and potashes seeking the big market to the south. Warehouses sprang up along the canal bank at Buffalo to receive the grain, lumber, rails, whiskey, fur and peltry to be forwarded to the seaboard, while boats coming through the canal from the east brought cargoes of salt, furniture, and general merchandise to the bustling Lake Erie port.

The Erie Canal fixed the destiny of New York City, forced it rapidly to preëminence as the national port of entry, and as the center of our export trade. It shifted the great trans-Allegheny route away from the Potomac, out of the belt of the slaveholding, agricultural South to the free, industrial North, and placed it at the back door of New England, whence poured westward a tide of Puritan emigrants, infusing elements of vigorous conscience and energy into all the northern zone of states from the Genesee River to the Missouri and Minnesota. The prairie lands which these new Westerners cultivated were, by means of the Lakes and the Erie Canal, made tributary to the growing metropolis at the mouth of the Hudson. New York became now commercially, as formerly it had been in a military sense, the keystone of the Atlantic shore arch. Baltimore, Philadelphia, and Boston lost much of their importance, and did not regain it even in part until railroads enabled them to reëstablish interior connections.

The local effects in central New York State were equally marked. The farmer found wheat quadrupled in price and the timber from his newly cleared land in reach of the market, for transportation from Buffalo to Albany declined in the twenty-six years after the opening of the canal from $88 to $5.98 a ton. But the farmer about Cleveland or Detroit also profited by these advantages. The consequence was that the price of land along the Erie Canal did not rise as rapidly as was anticipated, because the cheap and abundant prairie lands along the upper lakes were drawn like their products into the market, on the economic principle that every extension of transportation facilities tends to enlarge the area of competition.

However, there was compensation in the industrial impulse given by the abundant water-power developed by the canal and its feeders from the upland lakes. Mills and manufactories of great importance sprang up along the entire length of the New York state canals, as later on the Ohio canals. Especially pails, tubs, and woodenware were fashioned out of the adjacent forest by turning lathes run by water. The manufactured products found a market east and west by the Erie Canal. Water-power though cheap is local and immobile; hence it must command transportation facilities, or however perfect technically it is unavailable. Along the Erie Canal the water-power was one with the waterway, and for this reason reproduced in part the economico-geographical advantages of New England's early manufacturing towns which were situated along the line where river falls and ocean tide met.

Upon the West the Erie Canal had a marked effect, not only by carrying thither new accessions of population and bringing out its produce, but also by stimulating the western states to open up their interiors by canals in order to profit by the through waterway to the East. This influence was first felt in Ohio, whose long lake frontier on the north, narrow width between Lake Erie and the navigable course of the Ohio on the south and east, full-flowing interior rivers, and general location near to the western terminus of the canal, suggested the feasibility and advantages of an extensive system of artificial waterways, while its relatively dense population, due to its proximity and accessibility to old centers of settlement in the East, enabled the state to sustain the cost of such improvements. Soon after the work on the Erie Canal had been commenced in 1819, the subject came up before the Ohio legislature. Out of four possible routes —the Great Miami and Maumee valleys, the Sandusky and Scioto rivers, the Cuyahoga and Muskingum, and the Grand and Mahoning— two were chosen. Between 1825 and 1835 the Miami Canal (265 miles) was constructed, uniting Cincinnati with the Maumee, and the Ohio Canal (306 miles), connecting Portsmouth on the Ohio with Cleveland, by way of the Scioto, middle Muskingum, and Cuyahoga valleys. The effect upon the growth and prosperity of the state was enormous, especially of that large central area of prairie farms which had hitherto formed an isolated district, cut off from the highways both of the Ohio and the Lakes, but which now felt the stimulating effect of increased intercourse to the north and to the south. Cleveland and Toledo developed into active lake ports, holding a geographic position similar to that of New York, while Akron, Massillon, and other manufacturing points near the summit level of the canal flourished by reason of the accessible water-power, as did Rome, Syracuse, and Rochester in central New York.

The stimulating effects of the Erie Canal were felt still further west. Indiana was eager to take advantage of the low watershed in the northeastern part of her territory to unite the Wabash with the Maumee, and Lake Michigan with Lake Erie; but her plans of internal improvement were so extensive that the sparse population of a new state could not sustain the expense, and works half completed had to be abandoned.

The route of the fur-traders from Lake Michigan down the Illinois, which seemed peculiarly adapted to canalization, early (1816) attracted national attention as a military and commercial waterway between the Mississippi and Lake Michigan. The massacre at the mouth of the Chicago River in the War of 1812 emphasized the isolation of the northwest lakes. In 1817 Major Long in his report to Congress regarded such a canal as " first in importance of any in this quarter of the country ; " but though projected in 1825, owing to difficulties both technical and financial, it was not opened till 1848, and then proved disappointing as an investment, chiefly because the upper Illinois River needed artificial aid to make it navigable to any but small craft. The same difficulty was present in the waterway of Green Bay, Fox River, and the Wisconsin. The canal connecting the great bend of the Wisconsin with the Upper Fox was only two and one third miles long, but the sand-bars of the Wisconsin, the shallow, tortuous course of the Upper Fox, and the numerous rapids necessitating canals on the lower river, robbed this waterway of much of its value, and made it succumb rapidly to railroad competition. These western channels over the Great Lakes watershed, being constructed later (1848–53) than the Erie and Ohio canals, because on the frontier of settlement, felt the deadening grip of the railroad before their traffic had become established, and suffered moreover the disadvantage of reaching less navigable outlets into the Mississippi River.

The last extension of the Great Lakes waterway to

the west was the canal at the Sault Sainte Marie, which made Lake Superior accessible. The Erie Canal, by tapping Lake Erie above Niagara Falls, left to Canada the work of constructing the Welland Canal; but the eighteen feet ascent by lock from Lake Michigan to the level of Lake Superior was a natural task for the United States because of her longer coast-line and hence stronger interests in these more northern lakes. Commercial and military considerations led Canada to construct a similar channel on her side of the line, since Lake Superior is a link in the vast land and water route formed by the Atlantic, St. Lawrence River, Great Lakes, Canadian Pacific Railroad, and Pacific Ocean, which unites England with her colonies in Australia and the Orient.

The "Soo" Canal, opened in 1856, deepened to a ship canal in 1877, deepened again to twenty feet in 1896, developed its full importance only when it reduced the cost of transportation by admitting large vessels. Its consequences then were far-reaching. It brought about the transfer of the iron industry from the eastern to the western side of the Alleghenies. The iron ore from northern Michigan and Wisconsin is now carried at low freight rates to the manufacturing points in Illinois, Ohio, and western Pennsylvania, but a return westbound traffic in coal from the Erie ports fills the otherwise empty vessels, is carried therefore at the very low rate of thirty cents per ton to Duluth, and renders possible the manufacture of iron near the mines.[15] Since 1881 iron ore has formed about one half the tonnage through the canal.

While the " Soo " Canal has given great impulse to
the manufacturing interests along the southern borders
of Lakes Michigan and Erie, it has brought large accession of population to the Lake Superior counties of
Michigan, Wisconsin, and Minnesota, stimulated the
growth of Lake Superior ports like Duluth, Superior
City, and Ashland, and occasioned the development of
railroads westward into the Red River valley, whence
comes the flour, wheat, and other grains which help
swell the freight of the " Soo " traffic, and make it
larger than that of the Suez Canal.

Rapidly increasing commerce has demanded larger
vessels constructed on the most approved pattern and
made of steel. The mines near the Lakes provided the
material for these and in turn furnished the cargo for
the craft when ready for service. The result has been
an enormous increase of our merchant marine on the
Lakes since the deepening of the " Soo " passage. In
1877 our lake tonnage was only one fifth of the total
on the Atlantic, Gulf, and Pacific coasts; in 1887 it
was less than one fourth; in 1901 the tonnage on
the two oceans, exclusive of the small additions coming to us by the acquisition of Hawaii and Porto Rico,
was 3,526,024 and that of the Northern Lakes was
1,706,294, or nearly half.[16] The tonnage constructed
on the Lakes in the fiscal year ending June, 1901, comprised very nearly one half the total constructed in the
United States, but nearly two thirds of the tonnage of
steel steamers.[17]

Canada and the United States are natural competitors for the commerce of the Great Lakes. Our north-

ern neighbor, by her system of deep canals, has completed fourteen foot navigation from Lake Superior to
Montreal, while the shallow Erie Canal, with its small
locks, excludes all but the smallest craft. Montreal suffers, however, from the dangerous navigation of the
Lower St. Lawrence, which imposes a very high rate of
insurance on vessels bound for Canadian ports. Hence
the Erie Canal, despite its limitations, gets nearly two
thirds of the outgoing freight which seeks the seaboard
by water. However, the fact that a large part of the
traffic on these inland seas is limited to exchange between Superior and the southern Lakes, and the fact
that the raw material worked up in the big manufactures of Illinois, Ohio, and western Pennsylvania comes
out as finished commodities capable of bearing the heavy
charges of rail transportation to the seaboard, decrease
the amount of freight moving by canal to Atlantic
ports and act to the detriment both of New York and
Montreal.

New York is face to face with the problem of converting the Erie into a ship canal. The necessity is
urgent, for its old preëminence is passing from the great
port at the mouth of the Hudson. Philadelphia, Baltimore, Norfolk, and even New Orleans are gaining ominously, and the end is not yet. The latest report of
the New York commerce indicates not only a relative
but an absolute falling off. The warm discussion of
the proposed improvement of the Erie Canal in the
New York assembly of 1902 showed the importance
of the subject. The project to deepen the canal to nine
feet and enlarge its locks, or the other proposition to

give it twelve foot navigation for barges of one thousand tons, though ambitious schemes, and, because of
the great debt to be incurred, disheartening even to the
commercial courage of the Empire State, are themselves
inadequate and point to the alternative of national initiative. This is justified from the geographical standpoint.

Eight states, embracing 416,360 square miles and a
population of 27,150,437, border upon the Great Lakes
and will profit by a ship canal outlet, which, moreover,
would draw upon the contiguous territory of Canada.
The chain of the Great Lakes is an interior sea, affording the largest system of deep water inland navigation
on the globe. This system, with a general direction
east and west, extends from tidewater on the St. Lawrence and, by the Erie Canal, from tidewater at New
York 1400 miles into the heart of the continent; from
its western extremity at the head of Lake Superior the
distance to the Pacific Ocean is only 1700 miles. It
bears a yearly traffic of 40,000,000 tons, carried in 3253
vessels, and it serves our iron industry and grain export, two of the most important economic interests of
the United States. The outlet from this system should
be a national enterprise.

The Erie Canal was constructed in the light of experience of the War of 1812. Our Lake shore remains
a frontier coast of a thousand miles, and across the zone
of water, which has become our busiest thoroughfare,
lies a foreign power. A war with England in view of
our present friendly relations seems improbable, but national friendships are fitful things. A deepened Erie

Canal, admitting the smaller vessels of the navy in time of war, would have a military importance similar in kind, though less in degree, to that of the future Isthmian Canal, or the proposed Mediterranean-Garonne Valley Canal of France, or the Baltic-Black of Russia. All mean the uniting of two separated water frontiers. In the case of the Great Lakes this connecting link is the more important because the St. Lawrence River and the Canadian ship canals are large enough to admit English vessels. From the military point of view, then, this improvement is a task for the national government.

The very greatness of the natural resources of the United States accelerated the rate at which these resources were exploited. Production finally outstripped distribution, in spite of river and canal. The rapid development of our railroads was a protest against the economic inadequacy of inland waterways. A traffic interrupted by ice in winter and drought in summer, retarded by slow passage through canal or tortuous river course, trammeled by the geographic control of this or that port of outlet regardless of the final destination of the cargo, would not answer the wants of a country which had reached advanced economic development. The palmy days of water transportation in the United States were in the period from 1820 to 1860, when steamboating had improved its facilities, canals had extended the system of inland navigation, and railroads had not yet begun seriously to invade the province of our waterways, because the general economic development of the country had not yet passed beyond the agricultural stage. The bulky but cheap products of forest and farm found their economically appropriate transportation in the low-priced water carriage of that period. But with our industrial development came costly manufactured articles of relatively small bulk and large value, which were able to bear the heavier expense of railroad transportation. Hence we find that railroads developed earliest in the industrial areas of this country and waterways retained their value longest in the agricultural and forest regions. Even to-day rivers maintain their original significance only in economically retarded agricultural regions, as in Arkansas and northern Louisiana, and in the undeveloped area of the Cumberland Plateau, where the government improvements on the Great Kanawha, Big Sandy, Licking, and Kentucky tributaries of the Ohio, the Caney and Obey branches of the Cumberland, and the head streams of the Tennessee have proved a boon to the country. Here a commerce of lumber, railroad ties, barrel staves, and shingles is followed quickly by cargoes of agricultural and mining products, and the appearance of the latter will be the signal for the invading railroad, which seeks always areas of intenser economic activity and the concomitant denser population. The decline in the commerce of the Mississippi and its chief tributaries has everywhere been more than counterbalanced by improvement of small affluents opening up new districts.[18]

NOTES TO CHAPTER XIII

1. Ratzel, Politische Geographie der Vereinigten Staaten, p. 19.
2. Ratzel, Anthropogeographie, p. 341. 1899.
3. E. C. Semple, Development of the Hanse Towns, Bull. Amer. Geog. Soc., vol. xxxi. no. 3.

4. Ratzel, Anthropogeographie, p. 342. 1899.
5. C. Key, Railway Development in Federated South Africa, Engineering Magazine, May, 1902.
   Statesman's Yearbook of 1903.
6. Ratzel, Politische Geographie der Vereinigten Staaten, p. 539.
7. Eleventh Census, Transportation, vol. ii. pp. 247 and 396, 397.
8. James Hall, The West, its Commerce and Navigation, pp. 166–194. 1848.
9. Eleventh Census, Transportation, vol. ii. p. 399.
10. Ibid. p. 397.
11. Ratzel, Anthropogeographie, p. 343. 1899.
12. Winsor, The Westward Movement, pp. 248–256.
13. Washington, Letter to Governor Harrison of Virginia.
14. American State Papers, Misc. vol. i. no. 250, Washington. 1834.
15. Fairlie, Economic Effects of the Ship Canals, Annals of the American Academy, January, 1898.
16. United States Report of Commission of Navigation, p. 10. 1901.
17. Ibid. p. 17.
18. Eleventh Census, Transportation, vol. ii. pp. 418, 419.

CHAPTER XIV

THE GEOGRAPHY OF THE CIVIL WAR

" CIVILIZATION is at bottom an economic fact," at top an ethical fact. Beneath the economic lie the geographical conditions, and these in the last analysis are factors in the formation of ethical standards. The question of slavery in the United States was primarily a question of climate and soil, a question of rich alluvial valley and fertile coast-land plain, with a warm, moist, enervating climate, versus rough mountain upland and glaciated prairie or coast, with a colder, harsher, but more bracing climate. The morale of the institution, like the right of secession, was long a mooted question, until New England, having discovered the economic unfitness of slave industry for her boulder-strewn soil, took the lead in the crusade against it. The South, by the same token of geographical conditions, but conditions favorable to the plantation system which alone made slave labor profitable, upheld the institution both on economic and moral grounds.

Various circumstances combined to postpone the South's discovery of the economic unfitness of her industrial system. The presence of slavery was a deterrent to white immigration into the southern states and encouraged emigration of their unmoneyed whites into the free states, or withdrawal apart into the mountain

region of the southern Appalachians, whitl
petition of the plantation method of prod
not pursue them.   Furthermore, the rap
expansion of the slave power, due to rapi
of the soil by the extensive agriculture characteristic of
slave industry and the consequent need of taking up
new lands, stimulated also by the increased demand for
cotton which followed the invention of the spinning-
jenny, maintained a sparsity of white population, which,
being unrelieved by new accessions from without,
reduced the competition of the whites among themselves, kept down the beneficent struggle for existence
tending to produce a white laboring class, and thus
barred the experiment of free labor on the cotton and
sugar plantations.

The South thus maintained beyond its time the lavish
economic conditions of a new land, namely, extensive
area and sparse population.   The relation of the planters
to their territory recalls the still more expansive and
therefore less economic hold of the fur-trading French
Canadians upon their wide country.   In both cases
climate and soil, in the South too generous, in the
Canadian North too niggardly, combined, moreover,
with geographic conditions favoring expansion across
the broad Mississippi valley, as along the Great Lakes
highway into the vast forest interior, brought about
similar economic results.   Hence the South found its
basis of political ascendency in its mere extent of territory, and its life and death struggle was made in the
United States Senate, while the denser population of
the northern states gave them control in the House.

The balanced admission of slave and free states into the Union became a regular feature of political maneuvers; but the North had more raw material in the form of government territory out of which to form states than had the South. And here, too, geographical conditions played a conspicuous part. The boundary between Pennsylvania and Maryland, running just below the fortieth parallel, was the original Mason and Dixon Line dividing the free and slave states; but by the Ordinance of 1787, forbidding slavery in all territory north of the Ohio, this river became the extension of the line of demarcation, which, therefore, following the southwest course of the stream to its mouth at approximately the thirty-seventh parallel, encroached three degrees southward beyond the original Mason and Dixon Line. The early settlements in southeastern Missouri along the Mississippi at Cape Girardeau and New Madrid, in the days of French and Spanish supremacy here, were located in reference to the Ohio as the great river route from the East. These settlements had to be included in the young state of Missouri as admitted in 1820, and also in the privileges, one of them the right of holding slaves, guaranteed to them by the treaty of the Louisiana Purchase. Hence the Missouri Compromise Line at 36° 30′ north latitude can be traced back in its origin to the Ohio.

This line was of vast importance to the North. In 1820 the possible slave area east of the Mississippi, increased by the slave state of Louisiana, was nearly 23 per cent. greater than the area devoted to free labor (512,695 square miles as opposed to 417,680 square miles), although the southward trend of the dividing line from 40° to 36° 30' north latitude deprived the slave power of a broad belt of territory.   In the triangular area of the Louisiana Purchase, however, with its apex at the Gulf, there was now space for only two other states south of this line, and the westernmost of these was soon devoted to Indian reservations.   The recession of the Rocky Mountain watershed farther and farther to the northwest increased so much the domain of the free states.   But the enormous acquisitions of territory following the annexation of Texas and the Mexican War in 1850 meant a possible extension of the slave power to the Pacific, according to the terms of the Missouri Compromise.   However, geographic factors came in to wipe out the purely artificial line of 36° 30' north latitude.   California, in consequence of its remote trans-Rocky location, had received no negro element and had been populated largely by emigrants from the northern states or by small farmers from the southern states.   Hence it declared as a free state in applying for entrance to the Union in 1850, though its southern portion fell within slave territory according to the Missouri Compromise.

The sectional feeling originating in differences of climate and soil was further accentuated by the lack of intercourse between the North and South.   The great natural routes of communication, the Potomac-Ohio and the Erie Canal-Great Lakes, as also the chief railroad lines of 1860, ran east and west.   Moreover, intercourse along these routes was particularly active because their real termini were to be found, not in

Savannah, Charleston, Norfolk, Baltimore, New York, or Boston, but still farther east across the Atlantic in the ports of western Europe. The coast trade between the two rival sections touched only the outer rim of the southern country. Even the north and south course of the Mississippi, strong natural bond as it was in the whole trans-Appalachian region, did not suffice to offset the East's hold upon the states of the old Northwest secured by the double line of the Ohio and the Great Lakes.

Nor were the uniting influences of these great routes confined within the boundaries of the free-soil states. Geography knows no rigid lines of demarcation, no sharp transitions. The geographical unity of a river valley tends to be reflected in the homogeneity of its population, a homogeneity not only of constituent race elements but also of institutions and ideals. Hence along the line of the Potomac, Ohio, and Missouri there arose the phenomenon of the " border states," common-wealths not merely located along the northern boundary of the slave territory, but forming in themselves a great zone of assimilation of the conflicting elements to the north and south of them, and showing the inner and outer friction which belongs to every frontier.

Kentucky and Missouri, by reason of soil, climate, their industrial system, and their geographical lines of intercourse with the Gulf country, were in sympathy with the South. But though a large part of their population had come from Virginia, North Carolina, and Maryland, it had been drawn to a great extent from the non-slaveholding yeomen of the Blue Ridge frontier who had moved west by the Wilderness Road. At the same time the Ohio drew from its sources in Pennsylvania and New York northern farmers and artisans, or outbound immigrants and settlers moving along the National Road to Wheeling; many of these it deposited along its southern shores or carried further west into Missouri, which was receiving similar accession also by way of the Great Lakes. These men of northern or yeoman descent were at best merely tolerant of slavery. When it came to a question of slavery or the Union, their answer was prompt and decisive.

After a brief civil conflict, Missouri was secured to the Federal side. In Kentucky the strong planter element in the Bluegrass, the aristocracy of the state, was balanced by this yeoman element, by the industrial and foreign residents within its northern border, and especially by the non-slaveholding population in the rough uplands of the Cumberland Plateau. Mountain economy found no place for the negro or plantation cultivation in those sterile hillside farms, pathless forests, and roadless valleys. "The dwellers in the limestone formation, where the soil was rich, gave heavy pro-slavery majorities, while those living on the poorer sandstone soils were generally anti-slavery. This geological distribution of politics was common throughout the South." [1] The whole region of the southern Appalachians had therefore no sympathy with the industrial system of the South; it shared, moreover, in contrast to the aristocratic social organization of the planter community, the democratic spirit characteristic of all mountain people, and likewise their conservatism, which holds to the established order.

Hence when the rupture came between the North and South, the mountains declared for the nation, and thus formed a barrier of disaffection through the center of the southern states, all the more serious because coincident with the physical obstruction of the uplifted ranges. But this area of Union sympathy, because of its poverty, isolation, and remoteness from the Federal base, remained ineffective as regards positive assistance to the operations of the Union armies until after the fall of Chattanooga, though its moral support was always strong. One of the Kentucky mountain counties enjoys the distinction of having furnished to the Federal army the largest quota of troops in proportion to its population of any county in the Union. The uplands of East Tennessee put in a vigorous protest against disunion when that state seceded, and soon after the outbreak of hostilities endeavored to withdraw from the commonwealth. The same thing was true in the mountains of western North Carolina, northern Georgia, Alabama, and even in the hill country of South Carolina. West Virginia, a wholly mountainous region, dominated, moreover, through its northward-flowing streams by the upper Ohio, split off from its parent state. The geographical line of cleavage is instructive. It tells of the natural antagonism between upland and lowland.

Long-standing conflict of interests between the tide-water and highland counties of old Virginia had prompted schemes of separation prior to 1850, and in that year a new constitution was drawn up to right the grievances of the western section. When the permanent break came in 1861, the first ordinance for the new state adopted at Wheeling included only the counties west of the Allegheny range proper, and drew the boundary line along the mountain crest. The Old Dominion maintained its hold to the south upon the wide, fertile Valley of Virginia and thereby upon the New River, a West Virginia stream; but to the north the younger commonwealth encroached over the proposed dividing ridge east to the Shenandoah or Great North Mountain and its northern extension, the Sleepy Creek range. When erected into a separate state by the bill of Congress, July 10, 1862, West Virginia thus extended only to the western rim of the Shenandoah Valley. This district was geographically a part of the Valley of Virginia, and like it was adapted to plantation culture; but the two counties Berkeley and Jefferson in the lower end of the Valley, by their outpost location on the bank of the Potomac, were the natural recipients of northern settlers coming down the great trough of the Appalachians. Therefore in the summer of 1863, when most of the Confederate sympathizers were off fighting in the South, these two counties by a vote of the resident population were admitted to West Virginia: hence the erratic eastern boundary of the new state.[2]

The topography of West Virginia made it the sole aggressively loyal state of the border. Kentucky's attitude of declared neutrality was logical in view of the even balance of northern and southern sympathizers within her boundaries, but futile in the light of geographical conditions. A state situated on the outskirts of the scene of the war, like Kansas or Michigan, might have maintained such a neutrality. Wedged in between the Confederacy and the Union, Kentucky stretched its great length east and west from the Appalachians to the Mississippi across the very threshold of the South. Traversing its territory, the Mississippi, Tennessee, and Cumberland rivers opened up parallel avenues for an invading army into the heart of the Confederacy, while Cumberland Gap and the great intermontane valley of East Tennessee afforded a protected highway from the borders of the Bluegrass country to northern Georgia. Kentucky therefore throughout its length presented a strategic area, which was of paramount importance to the South, moreover, because the Ohio River along its northern border, bridgeless and almost fordless, presented the one strong defensible line in the Mississippi valley. Hence both sides made a dash for the possession of the state, invaded its boundaries in spite of legislative proclamation, and forced it from its neutrality. The territory of Kentucky was added to the area of the Union; but the flower of its manhood marched across the border into Tennessee to join the standard of the South.

In Maryland also geographical location was the all powerful factor in determining the destiny of the state. Here the predominant tidewater area and corresponding planter population made the secessionist feeling very strong; but Maryland's location behind the Potomac made it a highway for the northern troops hurried forward to the defense of the national capital, and therefore insured its prompt occupation by the Federal forces. The security of Washington was not the only object to be gained. The possession of Maryland meant the possession of Chesapeake Bay, a protected sea route to the gaping estuaries of the Virginia rivers.

The adherence of West Virginia and Maryland to the Union left the Old Dominion state a political peninsula projecting into a sea of hostile territory on the west and north, while the waters of Chesapeake Bay on the east were controlled by the enemy's fleet. This exposed position in itself insured Virginia a large share of the conflicts. To this was added the fact that the apex of her peninsula commanded the Federal capital, while the base sheltered the head of the Confederacy. The country between the Potomac and the James, therefore, became one continuous battle-field in the defensive and aggressive operations of both armies throughout the war.

The campaigns in Virginia were controlled largely by three geographical features, — the north and south reach of Chesapeake Bay, which enabled Federal troops to be transported by sea to any point on the Virginia coast, and later, when command of the James and York rivers was secured, supplies to be carried up to the forces occupied in the operations against Richmond or Petersburg ; the southeast course of the Virginia rivers across the lines of advance or retreat ; and finally the all important Shenandoah Valley, which, as a protected highway for a northward-marching army, enabled a Confederate force to threaten Washington just as surely as Chesapeake Bay rendered Richmond's position unsafe. In almost every movement and counter-movement of Federal and Confederate army these three elements of bay, river, and mountain valley played their parts.

A line running through Greenville, Petersburg, Richmond, Hanover, Fredericksburg, Manassas, Fairfax, and Washington indicates with a fair degree of accuracy the "fall line" which divides tidewater from Piedmont in Virginia. East of this line, which marks also the shortest route between the two capitals, the country is low, swampy, and cut by numerous parallel rivers which made the movements of an invading army difficult. Hence, with the exception of those engagements resulting from the advance up the James and York rivers on Richmond from Chesapeake Bay, all the great conflicts on Virginia soil were west of this line of one hundred feet elevation. Moreover, they had a definite situation in reference to the transverse streams, every one of which, from the Rappahannock and Rapidan south to the little Chickahominy and the Appomattox, was a natural line of defense for Richmond.

These battles of Piedmont Virginia marked the line of invasion which geographical conditions finally forced upon the Federal commanders. McClellan, in the spring of 1862, trying to avoid the difficulties of an advance across the tidewater country, took his army by Chesapeake Bay up the York and the James against Richmond; but the failure of his campaign was due not so much to Lee's valiant defense of the Confederate capital as to the fact that Washington had been left exposed by the withdrawal of the forces to the southeast, whence they could not watch the danger lurking in the Shenandoah Valley. Therefore, when in the fall of

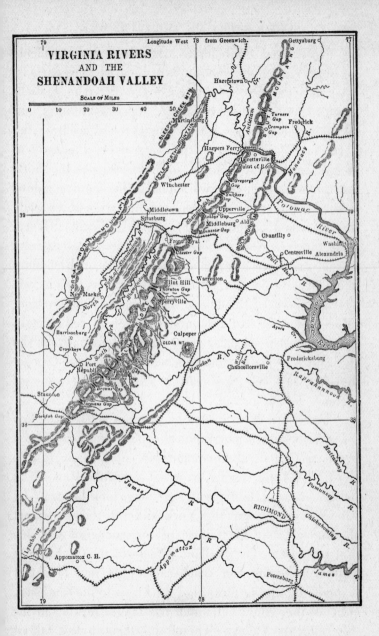

VIRGINIA RIVERS
AND THE
SHENANDOAH VALLEY

SCALE OF MILES

0   10   20   30   40   50

Longitude West 78 from Greenwich.

79   78   77

Gettysburg

Hagerstown

Martinsburg

Turners Gap
Frederick
Crampton Gap

Harpers Ferry

Lovettsville
Point of Rock

Winchester

Gregory's Gap

Snickers Gap

Upperville

Middletown

Strasburg

Ashby's Gap

Middleburg
Aldie

Manassas Gap

Front Royal

Chantilly

Chester Gap

Centreville
Washington
Alexandria

Warrenton

Flint Hill

Thornton Gap

New Market

North

Sperryville

Harrisonburg

Aquia Cr.

Culpeper

Crosskeys

CEDAR MT.

Port Republic

Fredericksburg

Staunton

Chancellorsville

Rappahannock R.

Rockfish Gap

Brown's Gap
Simmons Gap
Swift Run Gap
Fisher's Gap
Carman's Gap

James R.

Mattapony R.

RICHMOND

Pamunkey R.

Chickahominy R.

Lynchburg

Appomattox C. H.

Appomattox R.

Petersburg

James R.

39

38

BLUE RIDGE

GREAT NORTH MOUNTAIN

THE NORTH MOUNTAIN

SLEEPY CREEK MTN.

Potomac River

Bull Run

Rapidan R.

Monocacy R.

South Fork Shenandoah

North Fork Shenandoah

1862 a second advance on Richmond was proposed, McClellan's plan to go by sea again was rejected, and Burnside, who succeeded him in command, selected a route through central Virginia. The southward course of the Potomac was utilized for transportation to the great eastward bend at Acquia Creek just north of Fredericksburg on the Rappahannock, whose upper course from this point forms a natural extension westward of the line of the lower Potomac. The upper Rappahannock became therefore the Confederate line of defense. The concentration of Lee's forces at Fredericksburg, his entrenched position on the heights which mark the beginning of the Piedmont country, the difficulty and delay of the Federal advance across the Rappahannock by pontoons and ferry-boats while Confederate sharpshooters did their deadly work, all combined to bring about the disastrous repulse of the invading force.

The Rappahannock, about halfway between the two capitals, long remained the line of defense of both Federals and Confederates. Rifle-pits drew furrows along its banks, and pickets bristled at every ford and bridge. The southward course of the Potomac from Washington to Acquia and the possession of Maryland early gave northeast Virginia into the hands of the Federals; but the Rappahannock marked the limit where the Confederates could thus be taken in the east flank. Hence it saw the battle of Fredericksburg in '62, the bloody repulse again at Chancellorsville in '63, and the hard-won breach for the advancing Federals at the battles of the Wilderness in '64.

The line of the Rappahannock was especially advan-

tageous to the Confederates because its head waters led
up to the northern passes in the Blue Ridge which gave
access to the Shenandoah Valley, the third important
geographic factor in the operations in Virginia.   It is
the nature of mountains to lend themselves to strategy.
Their rocky ramparts conceal the movements of the
armies within their ranges and protect them from unexpected attack.   Their passes open the way for sudden
swoops down upon the enemy in the plain, and for
rapid retreat again, which may be covered by a small
force holding the narrow mountain gateway against the
pursuing enemy.   Mountains are always primarily barriers; passes, as the breaches in this barrier, are strategic points having a military and political history of
their own; the longitudinal valleys are the natural
highways, in general easy, along the axis of the upland
region within its guarding walls.   For example, the
Hindu Kush bars out expanding Russia from Afghanistan, but from their base in the plateau oasis of Merv
the Muscovite armies have moved southward up the
Murghab River to where its spreading head streams command the passes of the western Hindu Kush, much as the
sources of the Rappahannock and Rapidan ramify to a
series of gaps in the Blue Ridge.   Muscovite posts hold
also the parallel stream of the Tejend, and are creeping up
its course to the wide breach in the mountain wall known
as the " Gates of Herat;" for from this point the
upper stream, known as the Heri Rud, flowing down
a longitudinal valley of the Hindu Kush system, leads
eastward through a fertile country to easy passes which
carry the national highway of northern Afghanistan

over to the sources of the Cabul River, and thence down
this stream to the Khyber Pass and the teeming plains
of the Indus. Mountain, valley, and pass have figured
conspicuously for the past twenty-five years in the
British councils of India.

All of these phases came into play in the military
history of the Shenandoah Valley. As a part of the
Great Appalachian Valley, we have seen how, in the
period of colonial expansion, it received population from
the Great Valley in Pennsylvania, and passed on some
of its own elements to the valley of East Tennessee;
so now its military history had definite relations with
Hagerstown, Chambersburg, and Carlisle to the north,
Wytheville, Knoxville, and Chattanooga to the south.
Moreover, the Shenandoah Valley as a factor in the
campaigns of the Atlantic plain had a counterpart, with
only local variants, in the relation of the valley of East
Tennessee to the operations in the Mississippi basin.
In each case a mountain barrier and an isolated valley
highway flanked the movements in the adjacent lowlands. Cumberland Gap and the pass formed by the
Tennessee River at Chattanooga were the two openings
in the escarpment wall of the Cumberland Plateau giving access to the valley of East Tennessee, and hence
were the two strategic mountain points in the western
operations. But the numerous "wind gaps" which
notch the crest of the Blue Ridge every few miles made
the Shenandoah Valley much more accessible, enabled
the two armies in Virginia to play a game of hide-and-
seek in and out behind the barrier, and gave the Confederates opportunity for unexpected dashes through

these gateways either for attack or support, as in the case of Joseph Johnston's timely descent through Manassas Gap to reinforce the Southern troops at the first battle of Bull Run.

Napoleon said that Antwerp, as an outlet of the Rhine, was a pistol pointed at the heart of England. Now the deep groove of the Shenandoah was a pistol in the hands of the Confederacy, pointed at the heart of the Union, and from its barrel poured the deadly fire of Jackson's, Lee's, and Early's armies. The north-east and southwest trend of the valley brought its aim against Washington, Hagerstown, and Chambersburg, which therefore more than once felt its devastating fire. The only part of the Union territory east of the Alleghenies to suffer invasion was just this region in front of the Shenandoah opening. Hence it was always possible for the Confederate leaders to threaten Washington, and thus create a divertisement to draw off Federal forces from activities against Richmond, or retard the concentration of forces against a certain objective point until the season therefor had passed.[3] This was the purpose of Stonewall Jackson's raid from the Shenandoah in the spring of 1862.

The Federals, realizing the importance of the Shenandoah to the Confederacy, endeavored more than once to occupy it; but every advance up the Valley, by reason of its southwest trend, took them farther and farther from Richmond, and hence could have no other purpose than merely to clear this highway of the enemy. As soon as the Federal troops were withdrawn, the Confederates poured into it again by every gap. In the spring of 1862, while McClellan was advancing on Richmond, General Banks with a Union army had moved up the Shenandoah as far south as Harrisonburg, about twenty-four miles north of Staunton. Stonewall Jackson with two thousand men, to cut off Banks's retreat, struck through the Blue Ridge farther north and routed the Union force which had been placed at Front Royal to guard Manassas Gap. Banks succeeded in getting his forces north to Strasburg, and thence retreated down the Valley, with Jackson at his heels, across the Potomac. Union forces were now summoned to the Valley to drive out the intruders, — Fremont with his army from West Virginia and McDowell from Fredericksburg, on whose support McClellan had been counting for his operations against Richmond. Then came Jackson's retreat up the Valley, the battles of Cross Keys and Port Republic, his final escape through the Blue Ridge, and his return to the Confederate army at Richmond.

McClellan's campaign against Richmond, deprived of the necessary reinforcements which Jackson had been engaging in the Shenandoah, was thus a failure, and Lee saw his way clear for an advance on Maryland and the national capital. At the battle of Cedar Mountain, August 9, 1862, just north of the Rapidan, Stonewall Jackson dealt a heavy blow to General Banks, who with Sigel had brought his forces up the Valley from Winchester and Middletown, crossed the Blue Ridge by Luray Gap, and taken position at Sperryville on the upper course of Thornton River, a tributary of the Rappahannock. This position, just in front of the Luray and Thornton gaps, aimed by these passages to maintain a hold on the Shenandoah Valley and block any Confederate movement down this thoroughfare towards the north.[4]   At the same time it constituted the western end of the Federal line stretching along the Rappahannock to Fredericksburg and Acquia on the Potomac, and covering the approaches to Washington. The fierce battles on the old Bull Run ground, Centreville, and Chantilly at the close of August traced the route of the Confederate advance to the upper Potomac at Point of Rocks, where Lee's army crossed into Maryland. Here the Potomac is narrow and in general fordable; hence we find the Federal army later invading Virginia in this vicinity at Balls Bluff and Lovettsville. In a few days (September 10) Lee had occupied Hagerstown within the Great Valley of Maryland; but September 14 saw the Federal forces coming to the relief of Hagerstown, fighting their way through Crampton and Turner's gaps in South Mountain, the Maryland extension of the Blue Ridge; it saw Lee falling back to where Antietam Creek flows into the Potomac, to prevent McClellan's cutting off his retreat, while the next day Stonewall Jackson seized Harpers Ferry at the confluence of the Shenandoah and the Potomac, to secure the avenue of escape. The battle of Antietam, September 17, necessitated the withdrawal of the Confederates, but they found a safe route up the Shenandoah.

Lee's second invasion of the North in the summer of 1863, following the repulse of the Federals at Fredericksburg, was made also by the Shenandoah Valley.

He led his forces across the Blue Ridge by the passes
at the head of the Rappahannock and Rapidan, while
Ewell entered the Valley farther north by way of
Chester Gap and drove the enemy before him out of
Winchester, Martinsburg, and across the Potomac. In
the mean time Ashby and Snickers gaps in the Blue
Ridge, which were nearest the Federal forces massed
about Washington, were held by General Longstreet,
and General Stuart's cavalry was thrown out in front to
guard the approaches to the same. The engagements
at Aldie, Uppersville, and Middleburg were all battles in
the hill country to render secure the mountain passes
behind and protect Lee's advance down the Shenandoah Valley. Defensive warfare along a mountain
barrier, when vigorous enough, anticipates attack far
in front of the passes, and there deals its blow to the
invading enemy, like the victory of the Roman Marius
over the Teutons at Aquæ Sextiæ, or the Scotch victory at Bannockburn before the passes to the Highlands, or these Confederate successes before the gaps
of the Blue Ridge; or with a handful of men it de-
fends the pass itself, as the Greeks did at Thermopylæ,
the Moors at Roncesvalles, the Afghans at the Khyber
Pass against the advancing English, or the Confederates at Crampton and Turner's gaps in South Mountain.[5]

Thus every effort of the Federals to penetrate the
mountains being repulsed by Stuart's cavalry, the Confederate army advanced in security up the Great Appalachian Valley to Hagerstown, Chambersburg, Carlisle,
and to within a few miles of Harrisburg. But now the

numerous breaches in the eastern wall of the Great
Valley in Pennsylvania enabled the Federals to intercept Lee's lines of communication, just as before Jackson had been able to strike in and threaten Banks's line
of retreat. Therefore Lee turned and marched through
South Mountain to meet the advancing Federals at
Gettysburg. After his defeat there the Shenandoah
again afforded him a safe avenue of retreat, while
Meade with the Federal forces crossed the Potomac
near Lovettsville above the Point of Rocks and moved
southward to Warrenton, just north of the upper Rappahannock, whence he could watch the passes of the
Blue Ridge with the purpose of dealing Lee a final
blow as soon as he should show his head through the
mountain wall. It had been Lee's purpose to withdraw
to Loudoun County, Virginia, by way of Snickers and
Ashby gaps, but he was delayed by a sudden rise in the
Potomac which made it unfordable, and while his pon-
toon bridge was being prepared, Meade occupied these
passes. The close succession of openings saved him,
however. He made a feint of crowding his forces
through Manassas Gap, drew thither the bulk of the
Federal army, and then marching south rapidly to
Front Royal, moved out of the mountains by Chester
Gap to Culpeper, while Ewell's corps took the Thornton
Gap route a few miles further south and thus avoided
the delay of crowding the whole army through one
passway.

Thus the Shenandoah Valley had enabled the Confederate leader to accomplish his purpose of drawing
off the Army of the Potomac to the north, diverting

and diminishing the forces invading the coast of North Carolina and Virginia, and breaking up the plan of the enemy's campaign by the time the summer was ended.[6] Finally, when the last advance against Richmond was begun in the spring of 1864 and Grant settled down to the siege of Petersburg, history repeated itself in the Shenandoah Valley. Federal troops, as before under Banks, were sent up the Valley to prevent a possible demonstration against Washington, were defeated by the Confederates at New Market and driven out, rallied again under a new leader and advanced up the Valley as far as Staunton, and were threatening Lynchburg at the western end of the Confederate line; but being exposed to peril here in the heart of the enemy's country, they retreated over the mountains to West Virginia. Then, in the hope of making Grant loosen his hold on Petersburg, Lee tried the old expedient of 1862 and 1863. He ordered General Early with a strong Confederate force to sweep down the Shenandoah Valley, invade Maryland, and threaten Washington. Early came within gunshot of the capital, withdrew into the Valley with trains of plunder, moved northward into Pennsylvania, and once more made unfortunate Chambersburg feel the weight of the Confederate hand.

Then came the final chapter in the military history of the Shenandoah. Grant was not to be diverted from Petersburg. He sent General Phil Sheridan to clear the Valley. This meant not only driving out the Confederate force, but destroying everything which could furnish supplies to the enemy. Rich in its resources and strong in its loyalty to the Southern cause, the Shenandoah Valley had year after year given of its abundance to the armies of Virginia; but after Sheridan had done his work, this granary of the Confederacy was consumed, the marching and countermarching ceased, and the peace of desolation settled down upon the Valley.

Just as the Virginia rivers leading up to the passes of the Blue Ridge drew the lines of Federal and Confederate defense in the eastern operations of the Civil War, so in the trans-Allegheny country rivers and mountain passes determined the military lines. The campaigns in the Mississippi valley, though in general quite unconnected with those in the Atlantic plain, were of the nature of a vast flank movement upon the Confederates,[7] by which the southern end of the mountains was turned. Thus the Federals advancing from Georgia and South Carolina were enabled to take the Confederates in the rear, while another force, coming up the side path of the Great Valley from Tennessee, cut off Lee's retreat from Richmond by the southern passes of the Blue Ridge.

In no other war of history, perhaps, have rivers played so prominent a part as in the Rebellion. This importance is reflected in the names of the Federal armies, — the " Army of the Potomac," " Army of the James," " Army of the Cumberland," and " Army of the Tennessee." But while the Virginia streams, by their east and west course, simply drew the lines of defense adopted by either side, and with the exception of the lower James figured only by their fords, bridges, sudden rises cutting off advance or retreat, and head-stream paths to the mountain gaps, the western rivers, by reason of their large size, navigable character, and predominant north and south course, played the far more important rôle of highways for safe and easy water transportation right into the heart of the enemy's country. The great size of the territory constituting the western theater of war, with the consequent long lines of communication to be maintained by the invading Federals, made the western rivers more effective routes of communication than railroads, which could be easily torn up by one of the brilliant raids of the Confederate cavalry. The river routes required no watching, and the immense superiority of the North in steamboat building gave it all the equipment it could need.[8]

From the southern boundary of Illinois the Mississippi led through the Confederacy to the Gulf, while just to the east the Tennessee opened a highway to northern Alabama and the Cumberland into Tennessee. The parallel course of these three rivers therefore made the obvious line of advance for the Federals; and the Ohio, commanding the entrance to these three routes, was the natural objective of the Confederates. Hence the importance to the Southern cause of gaining the adherence of Kentucky, and the inevitable invasion of its neutral soil by both armies. A Confederate army under Polk entered Kentucky and secured the Mississippi by a strong position on the bluffs at Columbus near the Tennessee line, but Grant anticipated any further moves by seizing Paducah, which commanded the mouths of the Tennessee and the Cumberland.

Thus the Ohio line of defense was lost by the Confederates; the latter therefore drew their line from the Mississippi, where they strengthened the fortifications at Columbus, New Madrid, and Island No. 10, eastward near the Kentucky boundary across the Tennessee and Cumberland rivers at Forts Henry and Donelson to Cumberland Gap, which opened a route of communication with Virginia and, through the Valley of East Tennessee, with Georgia and Alabama.

The fall of Forts Henry and Donelson before Grant's combined land and naval force, and the defeat at Mill Spring, on the upper Cumberland, of the Confederates placed there to hold the connection with Cumberland Gap, compelled the Southern army to evacuate Nashville, abandon the line of the Cumberland, and fall back to their second line much farther south along the bend of the Tennessee. The vital points in this line were Corinth, the junction of two great railroads connecting the Mississippi and the Gulf with Virginia and the Carolinas, and Chattanooga, the southern gateway in the Appalachian wall. The highway of the Tennessee, immediately after the fall of Forts Henry and Donelson, brought the Federal transports and gunboats up the river to Pittsburg Landing, but the battle of Shiloh fought here retarded the proposed advance upon Corinth. This point was occupied in a few weeks, however, and then the battles of Iuka and Corinth along this line in the fall of 1862 and the battle of Stone River or Murfreesboro, Tennessee, represented the efforts of the Federals to advance east to the mountain gap at Chattanooga.

This point was of the utmost importance to the Confederates. It commanded the approaches to Atlanta
and the South Atlantic states, and controlled the intermontane bypath up the Great Appalachian Valley to
Virginia and central Kentucky.[9] Through this Shenandoah Valley of the West, the Confederates could
throw their invading armies over the Cumberland Mountains into central Kentucky by minor passes, eluding
the Federal forces at Cumberland Gap, threaten Frank-
fort and even Cincinnati, then retreat, driving their
captured horses and long wagon-trains of plunder from
the rich Bluegrass country, and vanish along the old
Wilderness Road into the mountains again, into whose
fastnesses few dared to pursue. Even if the invasion
was made by the plains across the middle Cumberland
River, as in the case of Bragg's raid in 1862, the Valley of East Tennessee was the safe line of retreat.
Thus it enabled the Confederates, so long as they held
Chattanooga, to make a safe flank movement against
the Federals in the Mississippi valley, just as the Shenandoah highway gave them this same advantage in Virginia and Maryland. But in compensation, the Tennessee River in the West, like Chesapeake Bay in the
East, gave the Federals easy lines of advance upon the
Confederate flanks.

The Federal invasion southward up the Tennessee
River was accompanied by a corresponding advance
down the Mississippi, because the wedge thus driven into
the enemy's territory along the smaller stream rendered
insecure or untenable their outlying positions on the
Mississippi. Moreover, certain physiographical pecu-

liarities of this river determined that the Confederate
withdrawal downstream should be measured by long
stretches, because only at rare intervals did its banks
present points which could be strongly fortified. The
river winds in gigantic meanders across its vast flood
plain, only now and then washing the base of its eastern bluffs, from the top of which its course could be
commanded by elevated shore batteries, but whither the
fire from passing gunboats could not penetrate. Such
points of natural defense were found only at Columbus,
Fort Pillow, Memphis, Vicksburg, Grand Gulf, and
Port Hudson; and here the Confederates concentrated
their efforts at fortification as the price of their hold
upon the river. One of these positions relinquished,
they had no choice but to fall back, no matter how far,
to the next similar point.[10] Only Island No. 10, by its
midstream position and geographical location near the
fortified point of New Madrid, gave them a secure base
of another kind whence to contest the passage of the
Federal flotillas.

When the loss of Fort Donelson and Fort Henry
left the Confederate position at Columbus like a remote
island in an enemy's sea, it had to be relinquished, and
New Madrid and Island No. 10 became the river outposts. When these fortifications also succumbed to a
combined attack of Federal infantry and fleet, the
Confederates dropped back to Fort Pillow, which in
turn was rendered untenable, owing to the Federal
occupation of Corinth far to the south, and had to
be abandoned for Memphis. When Memphis yielded
to the Federal fleet, the Mississippi was opened four

hundred miles down to Vicksburg; but this almost impregnable position on the bluffs two hundred feet above the Mississippi long resisted attack.

Meanwhile the Mississippi was being conquered upstream also from its mouth. After the control of the navigable course of the Tennessee was secured, the next problem of the war in the West was the command of the Mississippi, whereby this part of the Confederacy might be cut in twain, much as British control of the Hudson in the Revolution had divided the forces of the colonies. The coast blockade which was maintained along the whole seaboard of the southern states might thus be extended up the river highway on their western flank.[11] The fall of New Orleans before Farragut's fleet gave the key of the river into Federal hands; but the two Union fleets from Memphis and New Orleans failed to take Vicksburg, and the long stretch of river running through the enemy's country was difficult to control with this stronghold in the center unreduced. So the fleets fell back, the one northward to Helena and the other south to New Orleans. The Confederates seized their opportunity and fortified Port Hudson, a few miles above Baton Rouge but below the mouth of the Red River. The five hundred miles of the river now under their control between this point and Helena was to them the most important section. The Red River and the Arkansas brought them men and supplies from the trans-Mississippi states, which constituted the granary of the South, and gave them a secure line of communication through Texas and Mexico with Gulf ports.[12]

But lines were drawn tighter upon the Mississippi. Arkansas Post near the mouth of the Arkansas River was in a position to threaten the communications of Federal army and fleet operating from the north against Vicksburg; hence it had to be reduced. A land force of 30,000 troops and a flotilla proved just equal to the task, the control of the Arkansas River passed to the Federals, and this artery of the Confederacy was tied up. The Red River, however, the more important avenue to Confederate resources in the West, was safe so long as Vicksburg and Port Hudson, north and south of the river's mouth, were held by southern forces. The problem of the war in the West was the capture of Vicksburg. Its position was almost impregnable in consequence of the bluffs to north, west, and south; the only chance was to attack from the rear by difficult avenues of approach. But when Grant's generalship accomplished this maneuver and the gunboats were maintaining a blockade on the river, the fate of Vicksburg was sealed. Five days after the fall of this stronghold, Port Hudson also yielded, and the Mississippi was a Federal stream bisecting the western Confederacy.

The next move of the war aimed to divide the remainder of the Confederacy along the line of the Appalachians by securing Chattanooga, and thus controlling the lines of communication northward along the valley of East Tennessee with Virginia, and eastward around the southern foot of the mountains with the South Atlantic states. The battle of Murfreesboro was a conflict for the approaches to the Chattanooga gap, just as the battle of Mill Spring in southeastern Kentucky had for its prize the strategic point of Cumberland Gap. The battles of Chickamauga, Lookout Mountain, and Missionary Ridge in the corrugated uplands of northern Georgia gave this passway of the southern Appalachians to the northern armies. The Gulf states were cut off from the Atlantic states of the South, the theater of the war was now reduced to the Atlantic plain, and the national forces closed in around Richmond from the south as well as from the north. Stoneman's raid with a Federal army up the valley of East Tennessee from Knoxville, over the Great Smoky Mountains by the Watauga River, down to the Yadkin in North Carolina, and thence along the old pioneer line of communication to the New River and the Valley of Virginia, accomplished the destruction of bridges and railroads along the natural line of Lee's retreat from Richmond and hastened the surrender of the last Confederate army.[13]

The Civil War is characterized by an astonishing number of battles fought by combined land and naval forces. Gunboat and infantry charge united in all the great operations along the Cumberland, Tennessee, Mississippi, Arkansas, and Red rivers. The western streams were valuable not only as waterways for transports but as avenues of naval attack. The same phenomenon of combined military forces of land and water is to be noted in almost all the coast attacks. The seaboard cities of the southern states were protected by their prevailing location at the inner end of deep estuaries or bays, which were guarded by island fortifications at the entrances, while often a wide zone of swamp on the land side made difficult any approach from the rear. Hence the attack was usually made by the Federal fleet upon the outer island forts, which were reduced and straightway occupied by land forces from the northern transports. These then aided in maintaining the blockade of the city port in question, which, however, was rarely taken unless like Beaufort, North Carolina, it occupied an exposed position on the outer coast. Only Newbern at the head of the Neuse River estuary, Plymouth near the mouth of the Roanoke River, both in the Sound region of North Carolina, and New Orleans yielded to coast attacks in spite of their retired positions. In all these amphibious military operations we seem to see reflected the combination of land and water location in the places attacked.

To the political geographer, the success of the Federal side in the Civil War means the preservation of the large political territory. The evolution of political areas is marked by advance from the small to the large, from the city-state or local principality to the national kingdom or world-empire; it is characterized by increasing aggregation of territory, which minimizes the total amount of political boundaries, extends the area of fraternal feeling, and lessens the artificial barriers to commercial and social intercourse. From the standpoint of political geography, therefore, the disruption of the Union would have been a retrograde step, just as the defeat of the Transvaal means an advance even for the Transvaal itself, because of its absorption into a larger territorial body.

NOTES TO CHAPTER XIV

1. Shaler, History of Kentucky, p. 232.
2. V. A. Lewis, History of West Virginia, pp. 320–325, 373–375, 355, 386, 396. 1889.
3. F. V. Emerson, The Shenandoah Valley and the Civil War, Journal of School Geography, June, 1901.
4. War of the Rebellion, Official Records, series i. vol. xii. part ii. pp. 21, 22.
5. E. C. Semple, Mountain Passes, Bull. Amer. Geog. Society, nos. 2 and 3. 1901.
6. War of the Rebellion, Official Records, series i. vol. xxvii. part ii. pp. 302–305.
7. Fiske, Mississippi Valley in the Civil War, p. 1.
8. Ibid. pp. 107 and 194.
9. E. Kirke, Chattanooga the Southern Gateway of the Alleghanies, Harper's Magazine, April, 1887.
10. Fiske, Mississippi Valley in the Civil War, pp. 181, 182.
11. Ibid. p. 112.
12. Ibid. p. 183.
13. War of the Rebellion, Official Records, vol. xlix. part i. pp. 330–332.

CHAPTER XV

GEOGRAPHICAL DISTRIBUTION OF IMMIGRATION

In the expansion of the United States from a narrow
seaboard strip in 1783 to a broad continental territory,
three factors have been operating, — an abundant supply of free land due to continued acquisition of territory, a large foreign immigration, and the building
of railroads. These factors have been mutually interactive. The free land has attracted immigration, and
such additions to the population have increased the
pressure upon our political boundaries, causing them
to give way and then to be reconstructed far beyond
the original line. The railroads have opened up the
land we had and made it accessible to the throng of
settlers. But the most potent and persistent factor
has been always the presence of farm, field, and forest
to be had for the taking. This has stimulated natural increase of population, lured foreign settlers, and
encouraged the construction of far-reaching railroad
lines, while it has educated native and alien alike to the
large ideas of land-holding which have kept the people
spreading, till the expansion of the settled area from
decade to decade has almost kept pace with the growth
of population.[1]

The barrier of the Atlantic and the early geographical isolation of the American continent made a basis of

natural selection in the first colonists coming to this country. Immigration was voluntary except in the cases of slaves and the indented servants who were imported for the distinct economic purpose of supplying the labor much needed in a new land. Even in the opening years of the nineteenth century, the long voyage across the ocean in a sailing vessel and the relatively high cost of the passage were still deterrents to the indigent and thriftless; but in recent decades artificial influences of all kinds have stimulated and assisted emigration from Europe, while fast steamers have converted the ocean barrier into an open highway. The consequence is that in 1900 the whites of foreign birth and those of foreign parentage born in the United States comprised one third of our total population. In the North Atlantic division, which includes New England, New York, New Jersey, and Pennsylvania, these two elements represented very nearly 60 per cent. of the population, and in the Dakotas, Wisconsin, and Minnesota the proportion ran from 60 to 77 per cent.[2]

Great as has been the immigration into the United States since 1820, and steadily as it has increased, it has been only commensurate with the abundance of free land and the opportunities for work. By far the larger portion of these aliens have been unskilled laborers; but a new country with untouched resources, with forests to be cleared, land to be reclaimed, mines to be developed, and roads to be made, demands not so much skill as energy. Hence these crude foreign elements have served their purpose well, have contributed to the economic growth of the country, " quickened the pace of our development, and made us do things rapidly and on a large scale,"[3] while every advance of the frontier of settlement or expansion of our political boundary has subjected them more and more to the most American of American conditions, abundant land, and accelerated their evolution from European peasants to self-reliant, enterprising American citizens.

The land has been our great solvent. This steady European invasion of the United States has had no terrors for us, because our vast territory has enabled us to take in and assimilate. The powerful trans-Atlantic influences to which the country has thereby been subjected[4] have been diluted in their mere distribution over a wide area, weakened by their remoteness from their source, neutralized in part by each other; and all the time the 3,025,600 square miles of land and the free institutions of this oversea republic have been asserting the potency of the American environment. Those immigrants who have settled upon the farms have rapidly assimilated American standards of life, political, social, and economic, even where, as in some Norwegian centers in Minnesota, they have retained their own language; but those who, like the Russians and Italians, have crowded into cities, have been less Americanized, have transferred a bit of Europe to United States soil, because in the congested districts of a New York or Chicago the immediate influence of our continental area is lost.

The most striking fact brought out by the map showing the geographical distribution of our foreign population in 1890 is the paucity of alien elements in the southern states, due to the presence of that most

alien of all aliens, the negro. In this regard, the South
presents a sharp contrast to the remainder of the country. The great bulk of the old slave territory is
almost lacking in foreign elements. A narrow hem of
immigrants along the Gulf coast, an occasional spot
along the seaboard of Georgia and South Carolina, and
a ragged band of varying width within the old slave
periphery defined by the Ohio and the Mason and
Dixon Line, bear witness to the accessibility of all
coast regions to their oversea neighbors and the character of every land frontier as a zone of assimilation.
To counterbalance this foreign element in the frontier
portions of the old " border states," the free states just
north of the line show the effect of contact with the
old slave territory in the considerable negro population
of their southern counties. Maryland, with the largest
percentage (7.9 per cent.) of foreign born of all the
old slave states in 1900,[5] has a long frontier of contact
with the old free states and an extensive coast on the
Atlantic. Texas, with 5.8 per cent., and Florida, with
3.7 per cent., are young states with large unexploited
resources inviting immigration, and are accessible from
the Gulf. Louisiana, with 3.7 per cent., has a long
coast-line and the port of a great inland waterway; it
affords, moreover, a climate attractive to the Mediterranean peoples of Europe and a natural market for the
fruit trade, a form of commerce for which these people
show an appetency. In all the other southern states the
percentage of foreign born is insignificant. Only an
occasional mining center in the southern Appalachians
shows the presence of the immigrant laborer.

In the remaining portions of the United States where conditions of climate and soil permit settlement at all, the foreign born element is universally found. But here again the highest percentages are shown by the coast regions and especially by the vicinity of important harbors like the Puget Sound district, the lower Columbia, and San Francisco Bay on the Pacific, and all the seaboard of the New England and middle states on the Atlantic. Here speaks the accessibility of the coasts. All immigrants, except those crossing the borders from Canada and Mexico, must enter this country by the seaports. Many find their enterprise and money exhausted by the time they reach the gateways of the land. Such are therefore stranded in or near the ports, whence they gradually spread to the nearest cities; for the land thereabouts has long since been taken up. The cities alone offer numerous openings for the predominant unskilled labor of the immigrant, and also employment in their particular trades for the skilled artisans.

Hence the region of the coast and the region of the large cities show a high percentage of foreign born. The states of the North Atlantic division, which have passed from the agricultural to the industrial stage of development and therefore show great concentration of population in urban centers, and which moreover are nearest to the teeming sources of our immigrant supply in Europe, show a higher percentage of foreign born than do the states of the Pacific coast. In Rhode Island 31.4 per cent. of the total population is of alien birth, in Massachusetts 30.2 per cent., Connecticut 26.2 per cent., New York 26.1 per cent., and New Jersey 22.9

per cent. New Hampshire, with its short coast-line and industrial centers restricted to the southeast portion of the state, has nevertheless 21.4 per cent. Maine, which in relation to the rest of New England is a backwoods state with only limited industrial development and hence few cities, has only 13.4 per cent. of its total population of foreign birth,[6] in spite of an extensive coast-line. But as we have seen before in the history of this commonwealth, other geographical conditions of an unfavorable character have neutralized the advantage of an excellent seaboard. If to these percentages is added the quota of the native population of foreign parents in each state, the figures are found in every case to be doubled. Thus these two elements combined comprise more than half (50.9 per cent.) of the population of the northern Atlantic states.[7] During the past decade this section has experienced a greater influx of foreigners than ever before, due to a decided change in the character of immigration. In recent years 50 per cent. of all the immigrants reaching the United States shores have come from Austria-Hungary (including Bohemia), Italy, Russia, and Poland.[8] These people, emanating from conditions of retarded economic development, recruit the lowest class of laborers. Hence the sweat-shops and cruder manufactures of the big cities, as also the mines and railroads of this section offer them the readiest opportunity to gain a livelihood.

Next to this concentration of immigration along the accessible region of the northeastern coast, we observe on the map an irregular inner zone of sparser foreign element, succeeded by an increasing proportion of for-

eign to native born from the interior towards the land
frontiers. This is especially noticeable in the north
central states, reaching from Ohio to Nebraska and from
Missouri to the Canadian border, though the phenomenon can be detected also in New England and New
York. Ohio shows a growing proportion of foreigner
to native from its southern to its northern boundary.
In the next three tiers of states toward the west, the
same increase from south to north is observed. The
scanty proportion of one to five per cent. in southern
Illinois and Missouri swells to thirty-five per cent. and
more in northern Michigan, Minnesota, and North
Dakota.

In the successive stages of our colonial and early
national development, we have noticed the stratification
of native and foreigner from the east towards the
west, the later comer in general being located farthest
towards the setting sun, on the abundant lands beyond
the outskirts of continuous settlement. But in 1840,
after the area of settlement defined on the north by the
border of Lake Erie and the southern extremities of
Lakes Huron and Michigan (42° N. L.) extended to the
margin of the arid belt,[9] the frontier lay no longer
towards the west but towards the north; and thither
turned the immigrant. Hence the outlying states along
the Canadian boundary show a very large proportion of
foreign born, — North Dakota 35.4 per cent., Minnesota
28.9 per cent., Wisconsin 24.9 per cent., and Michigan
22.4 per cent. And if we combine with these the native
Americans of foreign parentage, the figures run from 56.7
per cent. in Michigan to 77.1 per cent. in North Dakota.

In this region of the northwestern frontier we find particularly those nationalities who, like the Scandinavians, only recently began to come to this country, or who, like the Germans, have made especially large contributions to our population in the past few decades. The Scandinavians did not send any considerable numbers before the decade ending 1870, when they comprised 4.7 per cent. of the total immigration. This proportion rose to 10.8 per cent. in 1890, and fell again to 8.7 per cent. in 1900 with the general decrease of the arrivals from northern Europe and the increase from eastern and southern Europe. The Scandinavians, therefore, except such as tarried by the Atlantic seaboard, located well towards the northwest frontier, where in 1890 they constituted 42 per cent. of the total foreign born in North Dakota, 34.45 per cent. in South Dakota, 46.05 per cent. in Minnesota, 19.21 per cent. in Wisconsin, and over 20 per cent. in northern Illinois, in Iowa, and Nebraska. The Germans, who between 1860 and 1890 comprised from one fourth to one third of our total immigration, constitute from 27 per cent. to 54 per cent. of the aggregate foreign born throughout this area except in the far-away Dakotas, which are remoter from the great German center along the western shore of Lake Michigan. The Scandinavian center, representing a later immigration, lies farther west along the upper Mississippi.[10]

The average percentage of foreign born in these outlying states of the central West (21.5 per cent.) is approximately reproduced (20.7 per cent.) in the young states and territories between the Rocky Mountains and

the Pacific.   The general remoteness of the Cordilleran
district from the areas of older and denser settlement is
reflected clearly enough in these figures, but even more
plainly in the foreign born combined with the native
born of foreign parentage.   These two elements together constitute from forty to sixty per cent. of the
total population in all the far western states except
Oregon and New Mexico, both of which had a considerable native population — American and Spanish
respectively — in proportion to their arable soil, when
the gold fever stimulated trans-Rocky expansion in the
fifties.   The presence of the unprogressive " greaser "
in New Mexico, and Oregon's more limited geographical
advantages compared with nature's munificence to other
Pacific states, have discouraged immigration.

Owing to the predominant aridity of this western
country, the foreign elements have found their best
opportunities in the extensive mining regions of the
Rockies, the Great Basin, and the Sierra Nevada, where
at different points they constitute over thirty-four per
cent. of the population; but they are distributed also
among the few irrigable valleys of the interior, and
form a large continuous area of alien settlement along
the Pacific seaboard, on the fertile slopes of the Coast
Range, and in the long fruitful valleys just behind.
Here field, orchard, vineyard, and garden offer congenial employment to Italians, Swiss, and Chinese.

The foreign elements in trans-Rocky America, however, do not represent only the outermost strata of
European immigrants who have moved westward from
the Atlantic ports of the United States.   They include

a large number of aliens whom the Pacific highway has brought from its farther shores to our western coasts, and who are only scantily represented in the eastern sections of the country. Here, therefore, in the case of these trans-Pacific immigrants we find an inverted west-to-east stratification, densest in the accessible regions of the Pacific coast, growing sparser with some regularity in the successive tiers of states towards the interior, and disappearing almost wholly in the eastern Mississippi valley, but appearing again in scattered patches about the large cities of the Atlantic seaboard,[11] whither they have come by the long sea route around Cape Horn or via the Suez Canal and Europe.

China, Japan, Australia, and the Pacific islands have all, according to their ability, sent contributions to our population. China out of its abundance has given most. Immigration of the Celestials began soon after the discovery of gold in California. The first came in from near Hawaii, where they already formed a considerable part of the laboring class. Between 1848 and 1852 some ten thousand arrived, and in the latter year the number leaped to twenty thousand, but soon declined to a few thousand annually. They found their first employment in the placer mines, either as under-laborers or working abandoned claims for themselves. Later, as the country became more densely populated and the crude industries of the mining camp became diversified, the Chinese supplied the laboring class which the phenomenally rapid growth of California under peculiar geographical conditions had not permitted to develop among the American population.

Measured in terms of comfort, money, and time, California was nearer to China than to the Mississippi prior to 1869, when the first overland railroad was opened. The easy, cheap ocean voyage from Shanghai or Canton to San Francisco Bay made an open highway for the Mongolian coolie, while the long caravan journey from the Missouri, measured in months and leagues, the expense and hardship of such travel, and the barrier of mountain and desert, kept out the large influx of Americans from the eastern states which might have yielded a laboring class. Moreover, those who came found such wealth of opportunity in the mines, the agricultural resources of the free lands, and in commercial ventures, that every American became his own master, and employees did not exist as a class except among the Chinese. These sturdy workmen, therefore, formed the sub-stratum of rough labor in the new state; on them rested the burden of cutting down forests, opening roads, reclaiming rich swamp lands, harvesting crops, acting as domestic servants, and in general developing the abundant resources of the young Pacific commonwealths.[12]

The Pacific as a geographic factor in the history of Chinese immigration has not operated with its full power, owing both to restrictive legislation in this country, beginning in 1855 and culminating in 1882 in the Chinese Exclusion Bill, and also to peculiarities of the Chinese themselves, the absence among them of Chinese women and the family life, and the annual return of large numbers to their original home. Their immigration was first curtailed, then arrested, while the

Chinese population which had effected an entrance before the barriers were raised has lacked natural increase and has been regularly depleted by repatriation. Thus the factor of the Pacific highway has been steadily weakened so far as Chinese immigrants are concerned, and with a few more extensions of the Exclusion Bill it will be nullified.

The returns of the Twelfth Census show a total of 89,863 Chinese in the United States proper as against 107,488 in 1890. Their distribution illustrates their relation to the Pacific. Over 75 per cent. of them are located in the trans-Rocky states and territories, and the heaviest proportions are found in the seaboard commonwealths. California heads the list with 45,753, which, however, represents a 35 per cent. decrease since 1890. Oregon follows with 10,397 and Washington with 3629, the smallness of the last figure being due to particularly violent local agitation against the unwelcome visitors. Outside the limits of the United States proper, Alaska, with over 3000 Chinese, and Hawaii, the great cross-roads station of the Pacific, with nearly 26,000, demonstrate the uniting influence of an ocean upon its bordering lands. The second tier of states from the Pacific, including Arizona, Nevada, Idaho, and also Montana, which owing to its close approach to Oregon properly belongs to this group, have only from 1300 to 1700 each; while in the next tier to the east the figures drop to 300 and 600,[13] thus showing the attenuation of the geographic influence of the Pacific in proportion to the distance from its coast.

The distribution of the 24,326 Japanese in the

United States in 1900 illustrates the same principle. Of this total 96 per cent. are found in the western division, where the largest numbers are located near the seaboard in California (10,151), Washington (5617), and Oregon (2501). But these figures sink into insignificance beside the 61,111 Japanese in Hawaii [14] California in 1890 claimed also nearly one third of the 5984 Australians settled in the United States. Our immigrants from the Hawaiian group and the other Pacific islands are distributed chiefly in the Pacific coast states. Here not only relative proximity but doubtless also similarity of climate has influenced the choice of the home-seekers. Many of the Pacific Islanders, owing to their mid-ocean position, have found it almost as easy to reach New York and Massachusetts; but Hawaii, which is only five days distant by steamer from San Francisco, has felt all the attracting power of California and the western states. [15]

A frontier, as we have seen, is always a zone of assimilation, but the intensity of the assimilation varies under different geographic conditions. A sea frontier, in consequence of the accessibility of the average coast to all the ships of the world, shows a degree of ethnic mixture not attained on a land frontier. China, which in consequence of geographic conditions and of national characteristics engendered largely by these conditions has almost no aliens in its vast interior, shows a considerable representation of the leading commercial nations of Europe and America in its great seaports. The delta region of ancient Egypt was the first part of that peculiarly isolated country ever to see foreigners

peacefully settling within its boundaries. That was
when under Psammetichus I. the Greeks settled in
Naucratis and Sais. Later the Nile delta became the
cosmopolitan area of the Mediterranean, while the ethnic
simplicity of middle and upper Egypt was almost untouched. So to-day the streets of Alexandria and
Cairo echo many languages. The southern seaboard
of France has various Mediterranean elements in its
population, with an alarming amount of the Italian
constituent. Greater New York contains Russian,
Polish, Italian, German, and Chinese towns within its
city limits. Boston, New Orleans, and San Francisco
can show almost as multifarious ethnic elements. A
comparison of all the maps showing the distribution of
our foreign born according to their nationality,[16] proves
how varied is the mixture of population along our
Pacific and northern Atlantic coasts.

Along a land frontier, on the contrary, the assimilation is as a rule limited to the two contiguous elements.
The close proximity of the two areas, however, the
similarity of climatic and hence of economic conditions
which usually prevails, and the intercourse across the
boundary in spite of trade restrictions, all stimulate the
shifting of population back and forth across the line.
Any special advantage offered now by this, now by
that side, will be the signal for migration. The large
German population in the Baltic provinces of Russia
to-day records the Teutonic expansion under the Sword
Brothers eastward across the Niemen and Duna as far
as Lake Peipus in the thirteenth century, while Slav
place names in eastern Germany attest not only the

expansion of Prussia over Polish territory, but also an earlier Slav encroachment towards the west, when these people extended to the River Elbe. Lately persecuted Russian Jews and German-Russians, restive under the harsh Russification process which has been going on in the Baltic provinces, have sought refuge across the boundary in Germany. The different frontier zones in Switzerland are strongly assimilated in language and race to their French, German, and Italian neighbors.[17] So, too, the northern land frontier of the United States is in no small degree Canadian, as the southern is strongly Mexican.

There were 77,853 Mexicans in the United States in 1890. Of these, 74,766 were in the southern tier of states along the Mexican border, and, with few exceptions to be specially explained, the largest proportions were found in the southernmost counties in those states. Even these large figures take no account of the strong infusion of Mexican blood in this border population which originated in the days of Spanish and Mexican supremacy here, and has been augmented ever since by a steady tide of immigrants in sombrero and zarape from across the line. Of the 7164 Mexicans in California in 1890, over thirty-five per cent. were located in the two southern tiers of counties, the proportions increasing towards the coast about Los Angeles and San Diego.[18] The only other considerable area of these people was found about San Francisco Bay, where a little over two thousand were distributed in San Francisco, Santa Clara, and Alameda counties.[19] Here speaks the accessibility of a seaboard country which

draws not only from trans-oceanic sources of population, but also from the neighboring coasts of its own
continent and its own hemisphere; for movements of
population tend to follow the coastwise trade. Mexico
therefore contributes inhabitants both to the sea and
the land frontier of California.

The 11,534 Mexicans in Arizona in 1890 were found
chiefly in the two southern tiers of counties, and New
Mexico's 4504 were located chiefly in Donna Anna
and Grant counties in the southwest corner of the state,
and in Lincoln County, also on the frontier. A line
of Mexican element extended northward from these
into Bernalillo County, which forms one of the rural
areas contiguous to Santa Fé, then east into San Miguel
County about the old Spanish town of Las Vegas, and
north across the Colorado boundary along the old line
of communication, now the route of the piedmont railroad, as far as Denver. The old Mexican centers and
the old routes have still the power to attract.

Texas counted 51,559 Mexicans among its citizens
in 1890. In all the counties along the Rio Grande,
except those arid sections almost without inhabitants,
they were found in numbers varying from one to seven
thousand, and constituted from twenty-seven per cent.
to fifty-five per cent. of the total population. They were
distributed in considerable numbers also in the second
tier of counties and in the whole southeastern corner
of the state, whither they came by sea as well as land.
The old Spanish administrative center of San Antonio
in the interior of the state has also become a gathering
place for Mexicans, who are strongly represented in

Bexar and the adjoining counties.[20]   But otherwise the Mexicans in Texas are a conspicuous phenomenon only along the land frontier.

The zone of assimilation characteristic of every political boundary is further illustrated in the case of our Mexican frontier by the incursions into United States territory made for many years by Mexican outlaws and Indians to steal cattle and horses, by the hot pursuit of the tireless Texas Rangers down to the Rio Grande and across the line, and the unrecorded battles or rope-end administration of American justice on Mexican soil. In the stirring days of the seventies and early eighties the security of the border settlements depended upon the sleepless vigilance of the Rangers, their skill in trailing the marauders through the brush of southwestern Texas and over the plains of Mexico, the tenacity of purpose which kept them days and nights in the saddle on a dash of three or four hundred miles, and their grim disregard of international law which put a certain finality upon their accomplished tasks.

Our other southern neighbors, though not in immediate contact with our frontier, have contributed elements to our population; and their distribution is interesting as illustrating the geographic control of location and climate, the accessibility of coast regions, and the tendency of population to follow the coastwise trade. In 1890 the census returns gave 23,256 immigrants from Cuba and the West Indies. The Gulf states claimed 13,300;[21] of these 12,282 fell to Florida alone, and 10,396 were distributed in the southernmost county of Monroe, which comprises also the Florida Keys, the stepping-stones to Cuba. Here they make up nearly the total foreign population. They are strongly represented by 1313 individuals also in Hillsboro County, which encloses Tampa and Tampa Bay, the great southern port for commercial intercourse with Cuba.[22] This county contains also a goodly sprinkling of Spaniards, who have evidently drifted in here from the southern island. The attraction of a congenial climate, similar to that from which they came, and geographical proximity have drawn thither these Cubans and West Indians. The 7235 of them in the northern Atlantic states show the effect of the great seaports and ship lines of this section. Over one half of the number are grouped about New York, Staten Island, and the western end of Long Island. This Caribbean district sends only a small contingent to the Pacific coast states, owing to the barrier of the Panama Isthmus.

In the distribution of immigrants from Central and South America, on the other hand, our two opposite coasts are more nearly on a par; and the Gulf seaboard, owing to the marked eastward projection of Brazil, loses its advantage of proximity for all immigrants coming from points south of the Equator. For such the port of New York is quite as near as New Orleans. The very small number of immigrants from Central and South America in the United States reflect our undeveloped relations with the southern part of the Western Hemisphere, where our political ascendency has no parallel in our almost rudimentary commercial and social intercourse. Of the five thousand South Americans in the United States in 1890, about 30 per cent. were found in the Pacific coast states, chiefly along the sea-board, and 35.8 per cent. in the North Atlantic division, also to large extent in or near the leading ports.[23] The Gulf states had only about 10 per cent. The 1192 Central Americans in this country in 1890 showed the same general distribution. The presence of a considerable number of immigrants from Cuba, the West Indies, Mexico, Central and South America in Missouri, Illinois, and Ohio, states bordering on the Mississippi and the Ohio, shows the geographic control exerted by the great central river system in the distribution of these southern elements, and the repelling influence of a harsh climate farther north.

The influence of immediate contact along a frontier upon immigration from the neighboring land is more strikingly illustrated in our experience with Canada than with Mexico. In the north a longer frontier, bordering moreover upon the most densely populated belt of Canada, a greater similarity in race elements, and the close proximity of superior economic conditions, all combined to produce the great exodus from Canada into the United States which has been going on for the past forty years. We notice here two outgoing currents. One sets strongly from the coast regions of eastern Canada to our northern Atlantic states, where Canadian immigrants are densely distributed along the whole littoral from Passamaquoddy to Raritan Bay. Hither they have come in vast numbers from Newfoundland, Nova Scotia, New Brunswick, and French Canada to find work in the great manufacturing centers, especially in those of New England, where in 1890 they often formed from 20 to 60 per cent. of the foreign born population.[24] There were 207,601 in Massachusetts, 93,193 in New York, 52,076 in Maine, 46,321 in New Hampshire, 27,934 in Rhode Island, 25,004 in Vermont, and 21,231 in Connecticut.

But not all the British Americans in these states were deposited by this coastwise current. The northern borders of New York, Vermont, New Hampshire, and Maine make a long line of contact with the Canadian boundary, and have been the scene of that frontier ethnic assimilation which we have come to know. Owing to the extreme narrowness of the United States territory in this northeastern corner, moreover, some of these border Canadians doubtless found their way by land routes southward to the Atlantic coast and contributed to the strong representation there.

The map showing the distribution of Canadians in this country according to the Census of 1890 shows how our northern neighbors spread over into our territory all along the frontier from the St. Croix River to Puget Sound, the distribution growing sparser from the border towards the interior, except where it increases again along our northern Atlantic coast. This incoming tide from Canada has met and mingled with that larger west and north bound current of European immigrants from the Atlantic ports. Hence our Canadian border, in point of ethnic elements, forms an exception to the rule of land frontiers inasmuch as it presents a varied mixture of nationalities, unlike the simple Mexican and American constituents along our southern boundary.

The distribution of British Americans in the United

States is much denser along the eastern half of the
frontier than along the western stretch beyond the head
of Lake Superior. The eastern half is contiguous to
the older, more densely populated provinces, and hence
to more abundant sources of immigrants than is the
western; and from New York to northern Minnesota,
the Great Lakes have facilitated and guided this northern immigration. The deep indentation of Lake Michigan has placed a large colony of Canadians in Illinois,
in the very heart of the Northwest, and brought them
within easy reach of the middle Mississippi and Missouri. The southern rims of Lakes Ontario and Erie
have a strong Canadian population. The two peninsulas of Michigan, being enclosed by Lakes Huron, Michigan, and Superior, are Canadian across their whole
width, and there is only a slight paling of color from
east to west on the map to show the increase of American influence towards the interior. The whole line of
the St. Lawrence and the Great Lakes has helped this
stratification towards the west, bringing the wheat-fields
of Dakota and Minnesota and the forests of Michigan
near the centers of population in eastern Canada.

The proportion of Canadians to the total population
is greatest in the frontier states, and in this group is
greatest in those older states which, owing to geographic conditions, have still a large backwoods area,
like Maine, New Hampshire, Vermont, and the Adirondack region of New York, or in the newer states to the
west which have been recently reclaimed from the
wilderness. Such regions in their extensive forests
offer congenial employment to Canadian lumbermen,

practiced in the woods of Canada, or further west in the prairie lands provide farms for those of agricultural tastes. The first motive has located Canadians in the forests of Maine, the Adirondacks, Michigan, and Wisconsin; the second has distributed them in central Minnesota, in North Dakota along the Red River of the North, and in the rich grain lands of Iowa. In North Dakota they form 12.61 per cent. of the total population and 28.29 per cent. of the foreign born. They are more strongly represented than any other alien people in every section of Montana, except in the two great mining counties, Silverbow and Lewis and Clark, which have attracted large numbers of Germans; also in the one frontier county of Idaho; and in Washington they form nearly one fifth of the total foreign born, being distributed most densely along the eastern and southern shores of Puget Sound. One center of greater density is found immediately on the land frontier, and another about the head of the inland sea, showing the tendency to coastwise expansion.

This tendency is demonstrated also by a continuous band of Canadian elements along the whole Pacific coast south to the Mexican boundary and by the location of all the denser centers immediately on the seaboard or at points like Portland and the Willamette valley readily accessible therefrom.[25] The old pioneer highway down the great Pacific Valley has also contributed to this distribution along the Pacific slope. All these coast states offer in their farming, grazing, and timber lands congenial employment to Canadian settlers. The isolated area of greater density in northern California on Humboldt Bay is to be explained by the important lumber industry of Humboldt County and by the port of Eureka, the best along this whole coast between the Golden Gate and the mouth of the Columbia.

The culmination of Canadian immigration into the United States was reached in 1890. The accessions to our population from this source amounted to 59,309 between 1850 and 1860; and every decade thereafter they increased rapidly till the returns of the Eleventh Census recorded the number as 392,802. During the last decade they have dropped suddenly to 3064. Nor do these small figures tell the whole story; they have to be supplemented by certain statistics compiled by the Canadian government, which show that the United States is furnishing annually a larger proportion of the total number of immigrants into Canada than any other country. Over 12,000 American settlers crossed the line in 1900; the number rose to 17,987 in 1901, and to 24,099 in the fiscal year ending June 30, 1902, when the figures were larger than those from the whole of continental Europe.

The United States then has lost to its northern neighbor about 55,000 citizens in the past three years; and even these reports are not complete, for they do not include a large number of settlers who treked across the border in their own wagons. The sudden turning of the tide of migration is due to the exhaustion of our supply of free arable land. This fact marks a crisis in the relation of our population to our area. Abundance of free land gave the United States the distinguishing characteristic of a youthful country; but the fierce rush to the Cherokee strip in 1893 and the ten applicants for every one of the fourteen thousand homesteads opened for occupation in the Kiowa-Comanche district in 1901, the recent invasion of our Indian reservations by American cattlemen, and this efflux of American farmers across the northern border, all indicate that the United States has attained its majority. Canada is now the lusty junior of the North American continent. Its vast area gives it great capacity for assimilation. As the only country in the North Temperate Zone now offering free land to home-seekers, it is retaining its own citizens, drawing a stream of emigrants from the United States, and is destined to attract the great tide of agricultural immigrants from Europe which until now has flowed into our own western lands.[26] The progressive policy of a great Canadian statesman, moreover, is taking advantage of this tendency. Canadian immigration agents are at work in most of our states; wonderful exhibits of Canadian grasses and grains are made at our state and county fairs; posters illustrating the northward bend of the isothermal lines west of Lake Superior and the ameliorating effects of the warm " Chinook " winds are displayed in market-places and school-houses; and finally the transfer of the migrating American is made as simple, comfortable, and cheap as possible. The consequence is that thousands of settlers are crossing the border to choose homesteads from the abundant grazing and farming lands of the Canadian Northwest.

The American emigrants are mostly from the border

states of the Dakotas, Montana, Minnesota, Wisconsin,
and Michigan ; from interior regions of uncertain rainfall like Kansas and Nebraska ; and from some of the
older states like Ohio, Illinois, and Missouri, where land
has risen in value and will sell for a good sum which
can be reinvested in the cheaper land of Canada for the
rising generation.   Many of the emigrants are people
who had moved to the United States from the older
provinces of Canada.[27]   The number includes also a
goodly sprinkling of Germans, Swedes, and Norwegians
who have been in the United States long enough to
learn the principle of prairie farming, and hence are
a vast improvement over the raw immigrants from
Europe.   These people are drawn from that part of
our country which has received the largest portion of
the vigorous, moral Teutonic stock of Europe.   When
we yield them up, therefore, we lose our best.

Nor is this all.   Most of the male adults are practical
farmers.[28]   Their knowledge is their best capital; but
they carry with them household effects, farm stock,
and agricultural implements by the carload, while their
pockets are well lined with cash.   They go to take up
free homesteads by the quarter section and buy improved lands in the vicinity, till they own often as much
as four thousand acres.   The 1661 persons who went
from Nebraska in the year ending June 30, 1901, to
settle in Manitoba and the Northwest Territories took
with them 154 carloads of effects and $1,762,050 in
capital.   During the same period North Dakota lost
2203 of its citizens who took out with them 384 carloads of effects, and Minnesota 2060 persons of the

farming class, representing a cash capital of over two million dollars.[29] All the emigrants are described as belonging to an excellent class of citizens and as being people of comfortable if not abundant means.

Such people we can ill afford to lose. And yet these are just the people who are alert to new advantages and who will seek them without hesitation. Migration is in their blood. Their local attachment is small. Bred in the large opportunities of a new country, they desire the same advantages for their offspring. Their table of square measure runs in terms of quarter sections and square miles. They want big farms and ranches for themselves and their children, and thus are unwilling to adopt the cramped standards of size and the intensive agriculture of a denser population and an older civilization. The demand that faces us, therefore, is more arable land on American soil. The amount and character of this American emigration proves that the demand is not a fanciful one. In the light of these facts, the recently proposed scheme of a national system of irrigation in the arid states of the West, whereby millions of acres may be reclaimed for cultivation, is not premature. It may even be called urgent.

NOTES TO CHAPTER XV

1. Richmond Mayo-Smith, Emigration and Immigration, p. 56. New York, 1890.
2. Twelfth Census, Bulletin no. 103, Table 23.
3. Richmond Mayo-Smith, Emigration and Immigration, p. 62. New York, 1890.
4. Ratzel, Politische Geographie der Vereinigten Staaten, p. 355. 1893.
5. Twelfth Census, Bulletin no. 103, Table 23.
6. Ibid. Table 8.

7. Ibid. Table 23.

8. Ibid. pp. 13, 14.

9. Eleventh Census, Population, part i. map of distribution of population in 1840.

10. Ibid. pp. cxlvi. and cxlvii. maps 13 and 14.

11. Ibid. Table 16, p. 437.   Also Twelfth Census, Bulletin no. 103, Table 16, p. 23.

12. Richmoud Mayo-Smith, Emigration and Immigration, pp. 236–246.

13. Twelfth Census, Bulletin no. 103, p. 23, Table 16.

14. Ibid.

15. For statistical basis of these conclusions, see Eleventh Census, Population, part i. Table 32, p. 609.

16. Ibid. maps 10–15.

17. W. Z. Ripley, The Races of Europe, chap. xi.   1899.

18. Eleventh Census, Population, part i. Table 32, p. 606.

19. Ibid. Table 33, p. 612.

20. Ibid. Comparison of Table 4, p. 41, and Table 33, p. 660.

21. Ibid. Table 32, p. 606.

22. Ibid. Table 33, p. 615.

23. Ibid. p. cxxxvi. and Table 33.

24. Ibid. p. cliii.

25 Ibid. Table 33, California, Oregon, and Washington.

26. J. D. Whelpley, The Isolation of Canada, in the Atlantic Monthly, Aug. 1901.

27. Canadian Report of Immigration, part ii. pp. 146 and 175.   Ottawa, 1901.

28. Ibid. pp. 116, 138, and passim.

29. Ibid. pp. 158, 161, 154.

CHAPTER XVI

GEOGRAPHICAL DISTRIBUTION OF CITIES AND INDUSTRIES

A COUNTRY in the germ, like the human embryo, passes rapidly through all the lower phases of development before it evolves to the type of the parent stock. Such was the history of colonial America. The settlers who came to people this country brought their best capital in the elements of European civilization. As exponents of this civilization they represented the forces of heredity. What transformed them was their environment, always the most potent factor for a young growth. The wide surrounding wilderness necessitated a return to a primitive type of living in order to cope with primitive conditions, but it generated a new order of adaptability, which is the strongest guarantee of a higher development. The men who learned the law of the wilderness gained the secret of its mastery. For the products of English garden and farm they accepted the savage commissariat of game, maize, and berries; for the busy London shops, Indian barter; for carriage and cart on well-built road, the highway of the streams with swift moving canoe or the blazed trail through forest and glade for the noiseless tread of moccasined feet.

Their settlements reflected in part the needs of defense which dictated the isolated, inaccessible sites of towns in the militant past. Fear of Indian attack located their first homes on Roanoke, Jamestown, Manhattan, and New Orleans islands, like the ancient Mediterranean cities of Tyre, Alexandria, Syracuse, and the Ionian colonies off the Asia Minor coast; or it placed their stockaded forts on hills or eminences as best protected against unexpected assault, like the Tuscan towns crowning high spurs of the Apennines, or the walled "burg" or citadel which characterizes the mediæval cities of Germany. Danger of aggressions from rival colonies along the coast suggested a retired location at the inner end of long, narrow inlets or estuaries, according to the instructions given the Jamestown settlers, just as the same motive of security located Rome, Ephesus, Smyrna, Troy, and Athens well beyond reach of the Mediterranean pirates.

Such were the temporary concessions made to the militant conditions of their savage environment by the early settlers. In conflict with the moulding influence of this environment were the commercial needs and purposes which an industrial, commercial people had brought with them from their trans-Atlantic homes. Firearms and the strength of increasing numbers soon emancipated them from rigid geographic control in the selection of their village sites. Still encircled by their stockades, however, the settlements lined up along the rivers to the head of navigation in order to command the sole means of communication with the interior, while their accessibility to the coast guaranteed the trade with the mother country, the constant source of the commercial influence. At the meeting-points of sea and inland navigation grew up the large towns, those with the best harbors and the easiest, most extensive lines of communication with the back country gradually gaining preëminence. For a long time, however, there was slight distinction between the small harbor and the large entrepôt; differentiation had not yet gone so far. The evolution of seaports in this country has been marked by increase of size attended by decrease of number.[1]

The colonial cities of America were essentially commercial centers, not industrial. They were markets where the country's new products were exchanged for the manufactured wares of England. The colonies as a whole were still in the agricultural stage of development. Only in New England had the geographical conditions of an excessively glaciated surface, abundant water-power, and accessibility by sea to outside sources of raw materials, curtailed the agricultural period and introduced the industrial. The New England towns became therefore manufacturing as well as commercial centers, but the products of their artisan labor were limited in variety, simple in character as a rule, and formed but a small part of the total exchanges. New York, Philadelphia, Baltimore, and later New Orleans became the typical commercial ports of a highly productive agricultural country.

The industrial phase of city development began with the close of the Revolution and the effort to gain industrial as well as political autonomy. It advanced most rapidly in those parts of the country where, from conditions of soil and topography, arable land was limited in supply and soon felt the competition of other agricultural areas more favored by nature; where the labor thus released from the farm sought other outlets for its energy, and where the laboring class thus formed received largest accessions from European immigration; where an invigorating climate stimulated the human energy necessary for the sustained labor of manufacture; and where superior routes of communication with other parts of the country, whether by sea, river, lake, or land, facilitated exchanges between such industrial centers and the agricultural areas. These conditions are found in the northeastern part of the United States, shading off in intensity towards the south and west, increasing towards the Atlantic coast. Hence the accompanying cartogram shows that three fourths of the one hundred and sixty cities of the United States, each having a population of twenty-five thousand or more, fall within a zone lying between thirty-seven and forty-three and a half degrees north latitude, and stretching from the ocean on the east to the ninety-seventh meridian just beyond the Missouri River on the west.

It is impossible to separate the commercial and industrial aspects of a city because of the close inter-relation between the two; now one aspect, now the other predominates, or again they may be equally strong, each contributing its part to the city's growth. The area of greatest urban development in the United States is the area of most abundant elements of production — labor, capital, water-power, fuel, and raw material of all kinds — as also of most numerous and varied routes of communication, not only with different parts of this country but also with Europe. Hence commercial and

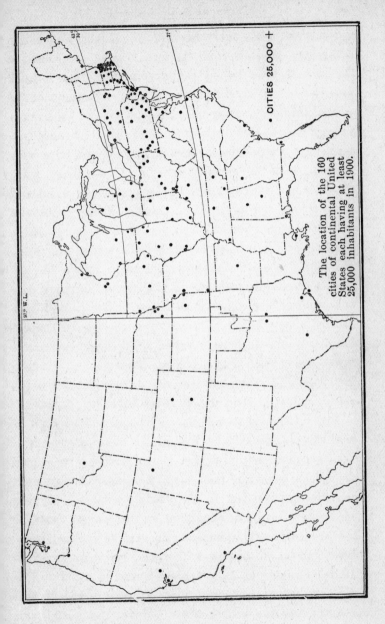

CITIES 25,000 •

The location of the 160
cities of continental United
States each having at least
25,000 inhabitants in 1900.

industrial activities are blended in all these urban centers in varying proportions which are determined largely by geographical conditions.

The modern city is essentially a point for collecting, producing, and distributing commodities of all kinds. Its location must be as accessible as possible. The site of the metropolis is at the cross-roads of the great world thoroughfares, where its markets can receive the products of all continents and all climes. All seaport towns are on the world's greatest and cheapest highway, the ocean. This is their first and greatest advantage. Their further development depends upon the area, fertility, and population of the back country which they command, and their means of communication with the same. The country for which Liverpool is the entrepôt is limited in extent, but vast in its content, in the amount and value of its products and in the size and demands of its population. Adelaide and Melbourne are good ports, but their growth will be limited by the desert character of the interior of Australia. New York and San Francisco are on opposite sides of the same country, command therefore the same area, and have equally good harbors; but while New York is connected by cheap waterways and almost level railroads with the interior, for San Francisco the high freight rates over the Rocky Mountains are prohibitive except for merchandise of small bulk and large value, and the arid plains and highlands of the West can never support the same density of population as the fertile region which lies within the range of attraction of the Hudson River port. Moreover San Francisco's trans-oceanic connections are longer than those of New York, and the outbound cargo to the Orient finds a far more restricted market than the highly progressive countries of Europe offer to the commerce of our Atlantic cities.

Coast cities develop because they are middlemen in the commerce of all the bordering continents. Cities located on a land frontier share in the trade only of two countries, their own and their neighbor's. Differences of geographical location, climate, soil, and degree of development tend to differentiate the products of the two communities and hence stimulate exchanges. Commerce between the two will depend upon the amount and kinds of their respective products. Interior cities are merely local distributing-points, whose commercial activity is determined by their command of routes of communication, and the contrast in economic development of the areas which they serve.

In the light of these general principles, let us examine the geographical distribution of urban centers in the United States. Of the twenty large cities having a population of 175,000 or more, nine (Boston, Providence, New York, Newark, Jersey City, Philadelphia, Baltimore, New Orleans, and San Francisco) are located on the coast, five more along the northern lake frontier (Buffalo, Cleveland, Detroit, Chicago, and Milwaukee), and five on the Mississippi and Ohio rivers (Pittsburg, Cincinnati, Louisville, St. Louis, and Minneapolis).[2] The twentieth city in this group is Washington, the national capital. Though many of these owe their size in part to their industries, nevertheless commercial facilities have made an important contribution to their growth.

The development of the large seaports in the United States reflects the advantages of a trans-oceanic market and of coastwise trade with other parts of the American continents. The active commerce which has long existed between New Orleans and the northern Atlantic states by sea would have been prohibited by the high freight charges of inland transportation, whether by rail or by river and canal. Four out of the six cities with a population of over 500,000 are located on the Atlantic coast, and the fifth, Chicago, is accessible from the ocean to sea-going vessels through the St. Lawrence River, the Canadian canals, and the Great Lakes. The same rule holds on the Pacific slope. West of the Rockies there are eight cities with a population of 30,000 or more, and six of these (San Francisco, Los Angeles, Oakland, Portland, Seattle, and Tacoma) are located on the seaboard; two (Salt Lake City and Spokane) in the interior. Though climatic conditions and the great fertility of the western coast region favor increase of population and hence of urban centers, nevertheless trade with the Yukon has been the making of Seattle, just as Asiatic commerce has stimulated the growth of San Francisco and Portland. Los Angeles will begin to exploit this field of wealth on a larger scale when the government improvements on the deep-sea harbor at San Pedro are completed.

The cities along the Great Lakes, though located on the inland periphery of the United States, yet enjoy the advantages of water transportation for a brisk interior coastwise trade with eight states of the Union and foreign trade with Canada. The most populous and productive provinces of Canada are located along the borders of these Lakes. Hence a considerable part of our large commerce ($157,585,695 in 1902[3]) with our northern neighbor moves across this Lake frontier and is forwarded by our Lake cities. The long chain of inland seas gives a wide range in the distribution of these exchanges. Chicago's markets in Canada stretch from Quebec to Port Arthur, and Canada lumber seeks the furniture factories at Chicago or Grand Rapids as readily as it does the wood-pulp mills of northern New York.

The cities on the Ohio and Mississippi are located on the outer margin of the urban industrial area and on the inner margin of the southern and western agricultural section. It is not a chance fact that St. Louis has developed the largest mercantile hardware house in the world. The city stands near the southwestern apex of the industrial peninsula, and the farming, mining, and grazing interests of the wide surrounding territory demand hardware more than any other one class of commodities. The two chief items of capital on a Texas or Nebraska ranch are so many heads of cattle, so many miles of wire fence. This same distributing function characterizes also in a less degree the ten cities with a population of over twenty-five thousand along the Mississippi, the seven on the Missouri, and seven on the Ohio.

An active commercial center necessarily possesses several of the qualifications of a successful manufacturing town. Situated on the focus of converging routes of communication, it commands abundant raw materials of various kinds and also an extensive market for the sale of its finished products. Its commercial interests insure the accumulation of capital necessary for industrial enterprises. Its labor supply, the last important

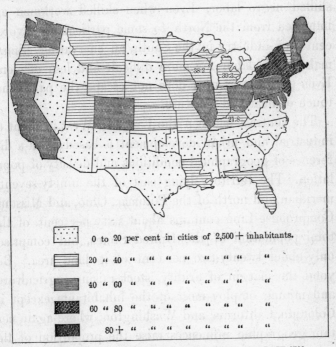

PER CENT. OF POPULATION IN CITIES HAVING AT LEAST 2500
INHABITANTS

factor in production, will depend upon the density and character of the surrounding population, and the conditions of climate under which its laborers must work. The urban development of a country is a fairly accurate index of its industrial progress. The accompanying cartogram [4] shows the striking contrast in point of city life between the North and South on the one hand and the East and West on the other. It is a climatic line (approximately 37° N. L.) which divides the urban North from the rural South, the area of abundant skilled white labor from the area of unreliable, unskilled negro labor. Even when skilled workmen are imported from the North to some particularly favored center of industry in the South, an enervating climate makes their labor far more difficult and less efficient. Even the entrepreneur does not escape the paralyzing touch of heat and moisture.

The contrast between the East and West in point of industrial and urban development is based upon a difference of geographical conditions and density of population. The territory lying east of the ninety-seventh meridian and north of the Potomac, Ohio, and Missouri Compromise Line contains about sixty per cent. of the total population of the United States, but comprises only about twenty-two per cent. of its land area.[5] Beyond the margin of aridity, stock-raising, agriculture, and mining employ most of the inhabitants except in Colorado, California, and Washington, where again certain geographic influences raise the proportion of the urban population. But in general, conditions of prevailing aridity limit the number of people the country can support, while the exclusion of all but costly railroad transportation across long stretches of desert and over mountain barriers discourages industrial progress and its concomitant urban development.

We distinguish three great areas of the United States in relation to cities: 1. The highly developed industrial area in New England and the middle states with predominant urban life (68.2 per cent.). 2. The Mississippi valley area with very successful agriculture and strong commercial and industrial centers in its northern half, where the urban population constitutes 38.5 per cent. of the whole.[6] 3. The great West, devoted chiefly to grazing and mining, with sparse population and a few cities as distributing-centers, some of them, however, combining industries closely related to their geographical conditions.

The area of intensest urban and industrial development is found in New England and the middle states. This development has been due to special local causes besides those general ones before enumerated as operating in the whole urban zone from the Atlantic to the Missouri River. The local advantage earliest felt was the water-power of New England and that developed in central New York with the construction of the Erie Canal. The whole Atlantic slope of the Appalachian Mountains naturally generates water-power. The rivers draining east cross an upland of resistant rock from which they descend to the coastal plain as cascades or rapids. This "fall line" stretches from the Delaware and Schuylkill southward as far as northern Georgia and Alabama, retreating farther and farther from the coast. In the lower courses of the New England streams, rapids and falls occur within a short distance of tidewater.

Besides this advantage of accessibility from the coast, as opposed to the long tortuous avenues of approach formed by the rivers of the most southern states, New England's water-power comes from concentrated falls

due to the predominance of hard metamorphic rocks
like granite, while in the middle and southern states
softer rocks give rise to shoals and rapids, often too
long and too wide to be utilized without heavy outlay.
This is especially true of the larger streams. The Susquehanna has very little power economically available.
In the southern states there are numerous abrupt falls
on the small head streams, which, however, are subject
to violent freshets due to the steep slope and heavy
rainfall. In New England and part of the middle
states, on the other hand, a more advantageously distributed rainfall, a larger precipitation in the summer
and autumn than in the winter and spring, compensates
for the evaporation during the heated term, and interior lakes of glacial origin regulate the flow of the
drainage streams all the year round. Hence water-
power was more readily available in the North Atlantic
states.[7] There it was developed earliest and most extensively, though it was brought into requisition also
for simple industries along the southern Piedmont. In
1870 it was used as much as steam in all the Atlantic
coast states. In New England it furnished 70.3 per
cent. of the total power used for manufacturing purposes, in the middle states 50 per cent., and in the
southern 49.8 per cent. of the total employed in these
different sections, though the water-power utilized in the
southern states amounted to less than half that employed
in either of the other sections.[8] This was doubtless a
result of retarded industrial development quite as much
as less favorable geographic conditions in the mountain
streams, as recent progress goes to prove.

Local water-power determined the beginning of Lewiston, Manchester, Lowell, Lawrence, Pawtucket, Waterbury, and Fall River in New England, and a line of towns along the Erie and Ohio canals. It has been an important factor in the localization of the textile industry, the manufacture of hosiery and knit goods, and of wood pulp and paper. The cotton industry has centralized chiefly in the water-power localities of New England and latterly in the Piedmont region of South Carolina, North Carolina, and Georgia. Though steam has become more important as an industrial force, nevertheless 32.6 per cent. of the total power used in New England for cotton manufacture comes from falls, and 34.8 per cent. in these three southern states.[9]

Abundant water-power, accessibility to the raw material, low cost of living, and peculiarly free conditions of labor (unrestricted employment of women and children, retarded development of labor unions) have brought about a rapid advance of cotton manufacture in the South, and started a migration of the industry from New England to the southern Piedmont. This has been a movement not only of the factory but also of capital. The water-power of the Piedmont region, estimated in 1880 at over 300,000 horse-power, though greatly developed in the past decade, will attain its full utility with its more general transmission in the form of electricity to desirable factory points. This has already been done successfully in Columbia, Pelzer, and Anderson, South Carolina. On account of the scattered distribution of this water-power, derived from relatively small streams as opposed to the concentration at Niagara and the Sault Sainte Marie, its remoteness from the commercial belt of the seaboard, and the recent period of its utilization, no cities of any considerable size have grown up with the industrial development to which it has contributed. In South Carolina, where this electrically transmitted energy has been most extensively employed in the South, and where the value of the cotton-mill products ranks second only to those of Massachusetts, urban manufactures nevertheless constitute less than forty per cent. of the total of the state reckoned in value of products.[10]

Power transmission and distribution by the use of an electric current has robbed water-power of that immobility which was its greatest disadvantage. It promises to give rise in the vicinity of large falls to urban centers of manufacture with an entirely new physiognomy, marked particularly by the lack of a congested factory district and by a predominance of the suburban character. The wide distribution of possible plant-sites, as opposed to the crowding of the mills about the falls as we see them in Minneapolis to-day, is limited only by facilities of transportation. And even the style of the buildings is modified, for massive walls to support heavy machinery are no longer necessary. Where smaller heads of water are used for the generation of electricity at commercially inaccessible points, as in the Sierras, the tendency is to concentrate this motive force at distant urban points in the lowlands, where raw materials and means of communication become determinants of the localization.

Water-power has been in many industries superseded

or reinforced by steam, but in paper and wood pulp manufacture it has remained supreme, constituting two thirds of all the power used for this purpose in 1900. The localization of this industry is instructive as showing the power of geographic control. The factors determining it are proximity to the supply of spruce and poplar, the timber chiefly used in making wood pulp, and found in its greatest abundance near our northern frontier and in Canada; water-power to operate cheaply the heavy grinding machinery; clear water for mixing pulp; and an adjacent market in the newspaper presses and publishing houses of the large northern cities.[11] Hence we find this industry localized along the Kennebec, Little Androscoggin, and Connecticut rivers; in Coos County in northern New Hampshire; in Saratoga, Washington, and Jefferson counties in New York at the outskirts of the Adirondack Mountains, where water-power is abundant and the local supply of wood can be supplemented readily from the Canadian forests by the St. Lawrence River and the Champlain Canal; and in Outagamie County, Wisconsin, where the abundant water-power north of Lake Winnebago and west of Green Bay, clear streams from the gravel-covered surface of an old glacial area, an almost inexhaustible forest of northern woods, and proximity to the large Lake cities, which are made economically nearer by cheap water transportation, offer all the conditions necessary for the successful conduct of this industry.

Where steam takes the place of water-power in the operation of manufactures, fuel becomes an all important factor in the localization of industries, especially where it is found near iron, the other raw material fundamental to industrial progress, or where the two can be readily brought into conjunction by cheap transportation. The anthracite coal-fields and iron-mines of the eastern Appalachians in Pennsylvania explain the numerous industrial towns of that section. These owe their origin or development to blast-furnaces, rolling-mills, and foundries. Canal communication with the lower Delaware and Hudson, freight cars moving for the greater part of the way on a downward grade from mines to seaboard have brought fuel and iron at low cost to New England and the middle states, and furnished the chief basis for the later industrial development of this section. Accessibility to the Pennsylvania mines explains the activity of iron and steel shipbuilding from the coast of Virginia to that of Maine, wherein the Chesapeake Bay and lower Delaware River districts contest the primacy.

The momentum of an early start and an extensive market have combined to keep eastern Pennsylvania an important area of iron and steel production; but rivals elsewhere have pushed it hard. The barrier of the Appalachians, as we have seen in a previous chapter, and the consequent higher freight charges across them, soon after the Revolution stimulated the exploitation of mining resources on the western slope of the mountains, and gave rise to industrial centers along the southern head streams of the Ohio. With the introduction of coke instead of anthracite as a fuel in blast-furnaces, and the increased use of ores from the Lake Superior mines, the center of the iron and steel indus-

try has migrated across the mountains to the Connelsville district, spreading from the sources of the Ohio
northwestward to Lake Erie, developing Johnstown,
Pittsburg, McKeesport, Duquesne, and Newcastle in
Pennsylvania, Youngstown, Cleveland, and numerous
other towns in Ohio. Connelsville grew up about its
coke-ovens, which furnish forty-eight per cent. of all
the coke manufactured in the United States. Ohio
owes its phenomenal industrial progress to its position
midway between the ore barges on Lake Erie coming
from Superior mines, and the coal arriving by canal,
river, and rail from the fields of western Pennsylvania
and West Virginia.

But this westward migration of the iron and steel
industry did not stop at Cleveland. Other Lake ports
were accessible to the northern ores and to Pennsylvania
coal and coke. Hence we find this industry contributing to the growth of Buffalo, Lorain, Detroit, Milwaukee, Chicago, and Joliet, the last getting its raw
materials by the Chicago drainage canal or short haul
on the railroad from Lake Michigan. The presence of
iron, steel, and fuel have stimulated in this vicinity the
establishment on a large scale of related manufactures,
such as iron pipes, machinery, engines, and locomotives.
Combined with a supply of hard woods from the adjacent northern forests, they have given rise to yet other
industries. The whole line of the Great Lakes has
seen a wonderful production of steel ships and barges
for the Lake trade, while a broad zone of cities between
the Lakes and the Ohio are making agricultural implements and farm-wagons for the grain belt near by.

The manufacture of the latter articles, which must stand heavy freight charges on account of bulk and weight, is largely localized in the state of Illinois, which makes 41.5 per cent. of the total, and in Ohio, Wisconsin, and Indiana. The industry migrates westward in the wake of expanding grain lands. This tendency is to be noticed in the falling off in value of the products in New York and Ohio during the last decade, and the increase in Illinois, Wisconsin, and Michigan.

The potency of iron ore and coal in giving rise to urban centers is demonstrated in the mineral belt at the foot of the Appalachian Mountains in northern Alabama, where the mere names of some of the recent towns and settlements — Birmingham, Sheffield, Bessemer, Irondale, Ironton, Dolomite, Newcastle, Connelsville, Carbonville, Coalberg, Coal City, and Coaldale — suggest the cause of their existence. Jefferson County, which comprises most of the far-famed Birmingham district, ranks fourth among the iron and steel producing counties in the United States. It is surpassed only by Allegheny County in Pennsylvania, (Mahoning County in eastern Ohio,) and Cook County (Chicago) in Illinois. This development has come in spite of adverse labor conditions, an enervating climate, and the lack of a cheap waterway to the sea. Yet Alabama pig iron is exported to England, continental Europe and Japan, and Alabama coal is beginning to compete with the Pittsburg product in the markets of New Orleans.

The whole of the Birmingham district is already dotted with industrial points, as yet only in the initial stages of their development, which promise to make it a second Allegheny County. Many lines of manufacture related to the iron and steel industry are already springing up.[12] Five years ago an old abandoned farm adjoining the blast-furnaces of the Tennessee Coal and Iron Company some six miles west of Birmingham was selected as the site of a new town named Ensley. In 1898 the streets had not been laid out, but were merely indicated by lines extending in some cases through fields which still showed the furrows of the plough. Now at the beginning of 1903 Ensley is an incorporated city with an estimated population of 10,350. It has one of the largest plants in the United States for the manufacture of wire nails and fencing, employing a thousand men; and a big establishment for the manufacture of steel cars is in process of construction.

Proximity to the abundant forests of the Appalachians to the north and the maritime pine belts to the south, and a location in the center of a vast non-industrial area will further insure a brilliant future to the Birmingham district. Its railroad facilities are excellent, but it needs water transportation to tidewater at Mobile to remove its chief handicap. Its destined industrial and urban development will more than justify the proposed canal, declared feasible at least along the Valley River to Bessemer, and the improvement of the Warrior River for six-foot navigation, whereby communication can be made southward through the Tombigbee and Mobile rivers to Mobile Bay.[13] It is claimed that such a waterway would reduce freight charges on iron eighty per cent.

All the industrial centers which we have considered

in relation to iron and coal production have had a close geographical connection with the Appalachian Mountains: they have been on the slopes of this upland area or have been readily accessible to it by rail, canal, and ocean on the east, or by the Great Lakes, rail, and Ohio River on the west. They may be called the cities of the iron belt, but some of them overlap the next or agricultural belt, and hence show industries sprung purely from the products of farm and field, just as some of the outermost towns of this zone, in turn, coincide with the few marginal manufacturing points of the grazing belt beyond. There is just one spot where all these belts overlap, where the industries emanating from the products of mine, forest, grain-field, and cattle ranch all come together, and that is Chicago. Hence its growth from a village of two hundred and fifty inhabitants in 1832 to a metropolis with a population of a million in 1890 and over a million and a half in 1900.

Ohio, Indiana, and Illinois, located in the great central wheat region, rank among the first four of the flour producing states, but they yield the primacy to Minnesota, which combines proximity to the spring wheat crop of the Red River valley in the United States and Canada with the water-power of the upper Mississippi. As flour and grist mills answer chiefly only local demand and do not belong by nature to city industries, we find them widely distributed; only those large establishments producing for a big domestic and foreign market contribute to urban development. Hence Minnesota's 14.1 per cent. of the country's total production gauged by value means a strong localization; and the fact that nearly five eighths of the state's product comes from the mills at the Falls of St. Anthony indicates the power of geographic control in specializing the industry and stimulating the development of Minneapolis. [See map showing productive areas of principal commercial staples in chapter XVII.]

Proximity to the great central grain-fields has favored the production of malt liquor in St. Paul, Milwaukee, Chicago, St. Louis, and Cincinnati, though a local demand from a large German population has also contributed to the result; and the same factor of location has been operative in the manufacture of distilled liquors in Illinois, Kentucky, Indiana, and Ohio, the four chief states in this industry, ranking in the order named. The increased production toward the margin of the western and southern agricultural area is noticeable. Distilled liquors, owing to their relatively small bulk and large value, can bear the cost of transportation far better than the raw materials from which they are made; hence the industry shows a strong tendency to localize near the grain supply.[14] It has been a marked factor in urban development only in the case of Peoria and Louisville. In Kentucky especially it is distributed in small establishments in the rural districts, generally wherever a large spring or stream of limestone water has furnished this desirable ingredient of the Bourbon product.

There has been a steady westward migration, following the advance of population, of all industries based upon cereal production; and these have been closely attended in their westward course by the allied industry of slaughtering and meat-packing. The herd and the cattle ranch naturally precede the corn and wheat field in the order of economic development. Hence we find that whereas flour and grist milling is most highly specialized on the Mississippi, slaughtering and meat-packing has its most striking localization farther west on the Missouri. Here, however, an arid climate has reinforced the effects of later settlement.

The historical geography of this industry is instructive. It began, as we have seen, in the early backwoods farms beyond the Alleghenies with the salting of hog products for export down the Mississippi to New Orleans and thence to the eastern seaboard cities. The Ohio, Illinois, and eastern Mississippi valleys were the chief stock-raising region of that day. Hogs especially ran wild in the "range" feeding on roots and nuts, or were fattened in the sty on corn. The salt came chiefly from the abundant mines on the Great Kanawha to the river towns. After the opening of the Ohio and Erie Canal, Cincinnati could command also the product of the Ontario mines in central New York; hence its supply of salt was assured. Its hams and bacon had a large market, particularly in the South.

The industry arose in Cincinnati in 1818 and had its chief center there till 1861-62, but numerous packing establishments sprang up in Columbus, Chillicothe, Circleville, and Hamilton, all of which were located on the Ohio canals; in several towns along the Ohio, notably Louisville, along the Wabash, Illinois, and Mississippi rivers; and in Chicago, where the industry began to develop in earnest only after 1850.[15] In 1862 the center migrated westward from Cincinnati to Chicago where it has remained ever since, though the most striking industrial specialization is found beyond on the Missouri.

The slaughtering and meat-packing business had come in the mean time to include the preparation of cattle and sheep for the market, and the ranching frontier had been pushed farther west. Sheep are now raised extensively in the upper Rocky Mountain states where pasturage is abundant, and cattle range the prairies of Texas, Kansas, Nebraska, and the Dakotas. Extended railroad systems and the introduction of refrigerator cars for the transportation of meat render it more economical to kill on the border of the cattle belt and save freight costs on waste product. First St. Louis, then Kansas City, Kansas, and finally South Omaha, Nebraska, developed as packing centers. Each of the last two surpassed St. Louis in 1900, but both together hardly equaled Chicago in the value of their products.[16] Chicago's location near the margin of the grazing lands to the west and the corn area just to the south, and her unsurpassed means of transportation, especially for the export trade, has assured her the primacy. An important part of the commercial activity of the city centers about her stockyards, packing establishments, and the dozen or more considerable industries which have developed to utilize the waste products of the slaughter houses. Some of these are the manufacture of soap, candles, glue, fertilizer, gelatin, brewer's isinglass, glycerine, ammonia, bone-meal, neat's-foot oil, felt, pepsin, and knife-handles.[17]

The strong localization of the slaughtering and pack-ing industry in the central West is indicated by the fact that the solid block of six states in the corn belt — Illinois, Kansas, Nebraska, Indiana, Missouri, and Iowa —in 1900 yielded together 77 per cent. of the total pro-duct of the United States reckoned in value. Illinois led with 40.1 per cent. of the total, a decline from the 46.3 per cent. of 1890, while Kansas advanced from 10.3 to 11 per cent., Nebraska from 5.5 to 10.2 per cent., and Missouri from 3.4 to 6 per cent. in the same decade. A line of cities on the Missouri River, fron-tier posts of the urban area, manifest a very intense specialization in this industry, for it yields to South Omaha 96.3 per cent. of the value of all its manufac-tured products, to Kansas City, Kansas, 88.4 per cent. and to St. Joseph 60 per cent.[18]

Each frontier belt tends to pass more or less rapidly through all the stages of economic evolution, from ranch to field and from field to factory. The rate of progress is determined mainly by the extent of the zone in ques-tion, the quantity and quality of its arable soil, and the geographic conditions regulating competition. In the pastoral and agricultural stages competition hastens, in the industrial it hinders incipient development. Beyond the Great Plains, whose corn and pasture lands support the packing industry of the central West, lies the vast territory between the Rocky Mountains and the Sierra Nevada, a region of scattered but productive agriculture where irrigation is possible, of wide cattle ranges, and abundant mines. It is a land of limited population, which the live-stock interest tends to distribute widely, but which the mines and intensive agriculture dependent upon irrigation canals tend to concentrate in certain small, favored areas, while wide stretches of desert and bare mountains are left wholly uninhabited. These facts seem to be the explanation of the rise in urban percentages in many of the far western states as seen in the cartogram; but the striking urban development of Colorado, California, and Washington is determined in great part also by certain other geographic conditions.

Colorado's cities and towns have naturally sprung up in the vicinity of its mines and of its upland water supply for irrigation. This state, in addition to gold, silver, copper, and lead, has also an abundant supply of iron and coal, the essentials of manufacture, found notably along the one hundred and fifth meridian at the eastern base of the mountains. Remoteness from the centers of manufactures east of the Mississippi, the lack of cheap water communication, and the prohibitive freight charges on the bulky and heavy commodities required by a mining and agricultural population not only in this state but also in the vast intermontane territory behind it, have combined to exclude competition and force the industrial development of Colorado. Denver's geographical location at the foot of the Rocky Mountains, near three of the natural passways to the Far West followed by the Union Pacific, the Denver and Rio Grande, and the Santa Fé railroads, has made it a distributing-point for a large hinterland. It has a population of 133,859. Just to the south Pueblo, with a similar location and a population of 28,157, and Leadville (population 12,455), within the mountains, have a predominant industrial development growing out of their mines. But the Colorado urban industries include also extensive smelting works, the manufacture of iron and steel, of foundry and machine products to answer the local demand; slaughtering and packing, due to the adjacent live-stock area; and flour and grist milling which avails itself of the abundant grain crops near by.

On the Pacific seaboard, also, remoteness from the eastern manufacturing centers and the consequent lack of competition early gave an impulse to industrial development. The presence of trans-oceanic markets on the populous coasts of Asia, and coastwise trade along the western shore of the American continents widened the market which inland was limited for most commodities by a mountain barrier. Abundant products of forest, ranch, field, orchard, and vineyard furnished many of the raw materials of manufacture. However, the lack of coal and iron and of the cheap and reliable labor developing with a dense population have proven great drawbacks, but have not sufficed to counterbalance certain local advantages. In Washington and Oregon abundant water-power, especially on the Columbia and Willamette rivers, has in part compensated for the lack of coal for many years. Recently local mines have produced fuel of only fair character. California is finding a valuable substitute for coal in the products of its extensive oil-fields in the south, and in the generation and transmission of electrical energy from the plunging streams of the Sierra Nevada in the region of abundant rainfall in the north. From the Colgate power-house on the North Yuba River in the Sierras northeast of

Sacramento, electricity is transmitted 218 miles to San Francisco. On its way it crosses sixteen of the most populous counties of the state, and side lines run the street-cars in Oakland, operate a flour-mill in Stockton, and various industries in Sacramento, Benicia, and San Jose.[19] In the state of Washington, electric lights, street railways, and motors in Seattle and Tacoma get their power from the Snoqualmie Falls some seventy miles away. It is proposed in the near future to provide all the Puget Sound cities with electricity in a similar manner, as their proximity to the steep slopes of the Cascade Range and the more abundant rainfall in this northern latitude simplify the problems of generation and of transmission.

The Pacific coast states, like Colorado, are industrial epitomes in consequence of their isolation; but the list of their leading manufactures marks them still a frontier country where land is abundant and the resources of forest, pasturage, and field are still ample. Lumber and timber industries, flour and grist milling, and slaughtering are prominent in all three states. In Washington and Oregon fish canning and preserving, which springs from the run of the salmon in the northern streams, is balanced by the canning of fruit and vegetables in California. Allied to these industries are planing-mills, paper and wood-pulp mills in the northern states, where, in consequence of a heavier rainfall and more recent exploitation, timber is most abundant, breweries and distilleries, tanneries and saddleries, and shipyards especially for the construction of wooden vessels. Car construction and repair-shops tell of the western termini of the great continental railroads. The preëminence of sugar and molasses refining among the industries of California points to the proximity of the Hawaiian Islands, whose raw sugar seeks San Francisco as the nearest port of entry into the United States.

All of these industries are due to abundant raw materials. The manufacture of foundry and machine products, on the other hand, in spite of the high prices of coal and iron, has developed to large proportions purely as a result of geographical isolation in response to a local demand for mining tools and machinery. This has been the case especially in California, though to no small degree also in Washington and Oregon. The year following the discovery of gold in California (1849) saw the first foundry established in that state; and in 1860 San Francisco alone had fourteen foundry and machine shops. The stimulant of a strong local demand has brought the industry now to such development that even in the face of existing disadvantages certain lines of its products not only control the local markets but also enter the markets of the world. The next step has been the rise of iron and steel shipbuilding, and the construction of the battle-ship Oregon.

The marked industrial character of the cities in the Pacific coast states is indicated by the fact that urban manufactures constitute from 59 to 66 per cent. of the total industries of this section. There is far less of the " neighborhood industry," production for local consumption, which characterizes all the states between the Sierras and the Missouri River except Colorado, and most of the region south of the Ohio River, and which reduces their proportion of urban manufacture to 30 per cent. or less. And yet it is safe to say that the industrial and hence urban development of the Pacific coast region will make strides in the future. Already the acquisition of the Philippines and an increased trade with the Orient is having its effect. This whole region is a base of supplies for the Yukon country, whose development is still in its infancy. Oil from southern California and coal from the south Alaskan fields promise to solve the problem of fuel. The opening of the Isthmian Canal will some day bring Alabama ores or pig iron to the Pacific ports; and though it will also break down the barrier of eastern competition, the mechanical skill which has accomplished so much under adverse conditions can be trusted to hold the market which it has preëmpted.

NOTES TO CHAPTER XVI.

1. Compare list of ports in Brownell's History of Immigration between 1818 and 1840.
2. Twelfth Census, Population, Table 6, p. 430.
3. Annual Review of the Foreign Commerce of the United States for 1902, Table, p. 4368.
4. Twelfth Census, Bulletin 149, p. 21.
5. Ibid. based upon tables 5, 6, 16, 17
6. Ibid. Table 22, p. 10.
7. For exhaustive discussion, see article by Prof. G. F. Swain in Tenth Census, Water-power of the United States, vol. i., Introduction. Also W. M. Davis, Physical Geography of Southern New England, Nat. Geog. Monographs, vol. i., no. 9 ; and Israel Russell, Rivers of North America.
8. Twelfth Census, Bulletin 247, pp. 23, 24.
9. Twelfth Census, Bulletin 244, p. 7.
10. Ibid. Bulletin 140, Table 5.

11. Twelfth Census, Bulletin 244, p. 17.
12. Ibid. Bulletin 117.
13. Doc. no. 88 of House of Representatives, 55th Congress. Report of 1898 entitled "Waterway between Warrior River and Five-Mile Creek, Alabama."
14. Twelfth Census, Bulletin 180, pp. 7, 21, 22.
15. Ibid. Bulletin 217, pp. 29, 30.
16. Ibid. Tables, p. 31.
17. Ibid. p. 33 and Bulletin 190, p. 13.
18. Ibid. Bulletin 244, Table lx. p. 21.
19. Ibid. Bulletin 247, p. 14.

CHAPTER XVII

THE GEOGRAPHICAL DISTRIBUTION OF RAILROADS

THE internal commerce of any country is based upon geographical conditions of size, location, zonal extent, and various topographical features tending to modify soil and climate. The larger a country, the greater in general are its chances of a diversified surface, contrast of climate, and hence the multiplicity of products which leads to exchanges. Large extent means wide climatic variation such as that between New England and Florida, Minnesota and Louisiana. China's location between the twentieth and fifty-fifth parallels gives it a basis for internal commerce similar to that of the United States between the twenty-fifth and forty-ninth. Mere zonal extent is not enough, however. Russia's long stretch from the fortieth to the seventieth parallel affords no advantages equal to those of China and the United States, because so much of the area falls within the polar and sub-polar region; yet in view of its monotonous lowland surface, only this great extent makes possible an internal Russian commerce between the forested north and the grain-fields and cattle pastures of the south. Everywhere along its southern border, from the Black to the Japan Sea, Russia is gradually enlarging its limits, with the result that it is diversifying its products, as the cotton-fields of Russian Turkestan

show, while it is striving to secure for the same an outlet
to the sea. In the widely distributed British Empire,
the distinction between internal and foreign commerce
is lost; but the enormous size of that commerce, however it may be classified, has its basis in the area and
the almost unlimited variety of geographical conditions
embraced within the British possessions.

No other continuous political area in the world, with
the possible exception of China, shows so many different geographical divisions distinguished by different
staple products as the United States. The accompanying cartogram makes six such divisions. We would add
an Appalachian section of coal, iron, and lumber, and
insert a purely live-stock section in the arid belt of the
Great Plains, between the ninety-ninth meridian and
the Rocky Mountains, which should be distinguished
from the Cordilleran area of live-stock and mining
products. These two with the Pacific coast section are
divided approximately by meridional lines. The Great
Plains section marks an area of aridity due to remoteness
from the Atlantic and Gulf winds. The Cordilleran
is also an area of scant rainfall due to the enclosing
mountains, but is also a mining region in consequence
of the outcropping of valuable strata in this area of
upheaval. In the Pacific coast division, proximity to
the ocean and exposure to its mild western winds have
also created another climatic section running across the
axis of the zones. East of the ninety-ninth meridian,
on the other hand, latitude exerts its full influence in
drawing the dividing lines between the spring wheat,
corn, and cotton sections. Only the uplifted belt of the

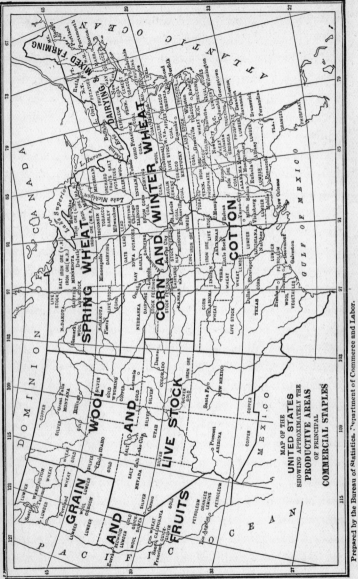

MAP OF THE
UNITED STATES
SHOWING APPROXIMATELY THE
PRODUCTIVE AREAS
OF PRINCIPAL
COMMERCIAL STAPLES

Prepared by the Bureau of Statistics, Department of Commerce and Labor.

Appalachians cuts across the grain of the parallels. The division formed by New York and New England is further distinguished as the area of excessive glaciation. This section, together with New Jersey and Pennsylvania, has reached most intensive industrial development, and is therefore dependent upon the rest of the country for subsistence and raw materials.

With this specialization of products, no section is self-sufficing and each depends upon all the rest. The stock of the Rocky Mountain states moves now to the packing-houses on the Missouri and Lake Michigan which are feeding New York, Boston, Halifax, and London; now to those on the Pacific coast which are furnishing meat for the growing Hawaiian, Alaskan, and Asiatic trade. The Columbia and Puget Sound regions send their lumber, canned fish, and wool to supply the East and get in exchange the finished merchandise which goes to swell the current of commerce. Cotton moves from the Gulf states to supply the spindles of New England, and coal from the Pennsylvania mines goes down to supply the plantations of the South. " The statistics of internal trade movements show that both the outgoing foreign and the domestic commerce of the country depend principally upon the industries engaged in the production of these staple products. These industries are known as the fundamental or extractive industries, and include agriculture, mining, stock-raising, lumbering, and fisheries. Manufactures and merchandise contribute a comparatively minor portion of the total tonnage carried in a given year. The dividend-earning tonnage of the western roads, for example, is found

chiefly in the volume of the grain and cattle shipments.
Lake transportation is supported primarily by ore, grain,
and coal freight. The southern railroads rely on cotton,
lumber, coal, and fruits for their prosperity." [2] Manufactures and merchandise constituted only about eighteen
per cent. of the total railroad tonnage originating on the
lines in the fiscal year ending 1899.

The vast scale of this geographical specialization of
products involves enormous movements of freight. It
has brought about the evolution of the American railroad system. Among the six sections laid down on the
cartogram, the most central position is held by the corn
and winter wheat division. This, as we saw in the last
chapter, is also the area of greatest urban and industrial
development, manufacturing activity decreasing from
the east towards the western limit. It is the area of
greatest concentration and distribution of products. It
contains eight out of the eleven leading grain markets
(Chicago, St. Louis, Toledo, Milwaukee, Kansas City,
Cincinnati, Peoria, and Detroit, named in the order of
their importance) of this country, and all of the four
or five most prominent stock centers, as we have seen.
It is the area of most extensive coal, iron, and steel
production, which with bread and meat constitute the
fundamentals in modern economic life. Hence in this
important central division the greatest movements of
internal commerce occur; it has become the channel of
practically all the great trunk lines reaching from the
interior to the Atlantic seaboard. This channel is
constricted on the north by the downward bend of
the Great Lakes and the geographical location of the

Mohawk depression, and on the south by the rough upland of the Cumberland Plateau, the determining line of the Ohio valley, and the central gaps of the Appalachian Mountains. Norfolk and Boston, together with all the intervening ports, are its eastern termini on the Atlantic coast. To the west, nine different railroad lines extend to the foot of the Rockies within the limits of the thirty-seventh and forty-third parallels, so that a railroad map of Kansas and Nebraska looks like a striped fabric. The meridional division of the sections of production in the Far West allows to the six overland railroads a wider north and south distribution, so that their course is determined chiefly by the best passes through the Rocky Mountains and the location of the few good ports on the Pacific. One road will ship eastward fresh fruit from southern California and wool from Arizona, while a more northern line will bring these same products from Oregon and Montana.

The great breadth of the United States and its highly favored interoceanic position has given to its railroads a predominant east-and-west direction. The expansion of the American people from seacoast to the interior, and from the interior on to the sea again, has been repeated in the history of American railroads, but at a tremendously accelerated rate. The star marking the center of population was poised on the summit of the Alleghenies in 1830 while infant railroads only here and there were crawling up the first easy grade of the Atlantic coastal plain. Twenty-one years later a line of rails stretched through central New York to Lake Erie, and the Baltimore and Ohio road had reached an equally

western point in Piedmont, West Virginia (79° W. L.),
close on the trail of the star, which was moving down
the sunset slope of the Allegheny Plateau towards Park-
ersburg and the Ohio. But eighteen years after that, an
overland express drew in to the shores of San Francisco Bay, while the star still lingered in the valley of
the upper Ohio.

Railroad development in the United States proceeded
with amazing rapidity, so that it soon outstripped that
in all other countries. The great size of our territory,
its long distances, made railroad communication more
necessary than in Europe. Its simple continental build,
yielding long stretches like the Atlantic plain, the
Mississippi valley, the prairies, and the Pacific valley,
was favorable to railroad construction because of the
relatively few obstacles which it presented.[3]   Moreover,
the oneness of our country, as opposed to the political
dismemberment of Europe, was conducive to the operation of great through lines and that consolidation of
interests which early appeared.

In the first decade of their history, these roads, which
in forty years were to rib the continent with steel, were
merely accessories to established routes of water travel ;
or they united some adjacent interior town with the seaboard ; or they served as a short cut across the land between two ports which had previously established active
intercourse by more devious coastwise communication.

Among those railroads whose objective was an established waterway, we find the short line running sixteen
miles from Carbondale to Honesdale (1828) and connecting the Lackawanna coal-mines with the Delaware

and Hudson Canal; the line of seventeen miles from
Albany to Schenectady, opened in 1831 to accommodate
passengers on the Erie Canal by saving them the long
détour of the waterway between these two points; the
road southward from Saratoga to Schenectady (21.5
miles) opened in 1832, and the branch off this line from
Ballston to the Hudson at Troy (25 miles), completed
in 1835; the road from the head of Lake Cayuga southward to navigation on the Susquehanna River (34 miles),
opened in 1834; the railways over the high land between Philadelphia and Columbia on the lower Susquehanna (81 miles), and the Portage Road over the summit
(2491 feet elevation) of the Allegheny Mountains (36
miles) just east of Johnstown, both opened in 1834, to
form links in the canal route joining Pittsburg and
Philadelphia; the Baltimore and Ohio, which in 1833,
after a piecemeal construction, extended to Harpers
Ferry (81 miles) on the Potomac, and there tapped the
canalled river whereby passengers avoided the long
détour of water travel around the end of the Potomac
peninsula; the short independent road from Harpers
Ferry up the Shenandoah Valley to Winchester (32
miles), opened in 1836; the little roads in Louisiana
from Port Hudson on the Mississippi (21.5 miles) northeast to Clinton, opened in 1833, from Bayou Sara in
the same vicinity to the town of Woodville (27.5 miles),
opened in 1842, and that from New Orleans running
eight miles to Carrollton, opened in 1835.[4]

All these roads, with the exception of two, are distinguished by very short length, which emphasizes their
merely auxiliary character. But the last one to be

mentioned in this class, the road between Boston and
Albany (1842), owing to the lack of inland waterways in New England, had to strike two hundred miles
across hill country and river valley to reach the growing western commerce which was coming out by the
Erie Canal. This road, therefore, owing to the side-
tracked position of Boston in relation to the great interior routes of the country, had to leave the convenient
level of the tidewater and lowland, and take a more
difficult upland course to the debouchment point of
the Erie Canal. Hence in point of length, which is the
measure of urgency, it resembles the Baltimore-Potomac
and the Philadelphia-Columbia lines; all three tap
established routes to the productive trans-Allegheny
country.

The second class of early railroads comprises those
aiming to connect inland points with the seaboard, which
commands the world's great waterway. They are
characterized as a whole by greater length than the first
class. Among them we find the germ of the Boston and
Albany, the road running from Boston to Worcester
(44 miles), opened in 1835 and extended to Springfield
in 1838; the line from Boston to Lowell (26 miles) in
1835; from New Haven to Hartford in 1839 and
extended along the easy route up the well-populated
Connecticut valley to Springfield in 1844, to Greenfield
in 1846, to South Vernon on the Vermont border in
1849, whence it branched off up the little Ashuelot
River to Keene, New Hampshire, in 1851; the road up
the Housatonic valley from Bridgeport, Connecticut,
forming a conjunction with the Boston and Albany

route at West Stockbridge on the New York state line in 1842; the Philadelphia and Reading road (98 miles) and a branch from Lancaster on the already established Columbia and Philadelphia line running off 54 miles to Harrisburg, both opened as early as 1838, and that in spite of competing canals; a line from Portsmouth at the mouth of the James River in Virginia running 80 miles back from the coast to Weldon, North Carolina, opened early in 1835; one from Charleston, South Carolina, to Hamburg on the Savannah River opposite Augusta, opened in 1833, when its 135 miles made it the longest continuous line in the world, and extended by a branch to Columbia in 1840, whereby the cotton of the interior could be more readily concentrated in the chief port of the southern Atlantic seaboard; and finally another road subserving a like purpose, opened in 1841 throughout the 192 miles between Savannah and Macon, Georgia.

The third class of early railroads was designed to connect coast points, and therefore, from Pamlico to the eastern end of Long Island Sound, had to compete with the inside coast passage of canal and bay and the highway of the open ocean. But the circuitous ocean route from point to point and the slow progress through the canals, winter storms on the outside passage and ice blockade on the inner channels, necessitated more direct and reliable communication. Hence we find the early development of what we may call the coast or " fall line " railroad. The dates of construction of this geographical line of roads, made piecemeal by different companies, as were all early ventures of this sort, are instructive as indicating the intensity of water competition, which varied with the directness of the sea route. The railroad from Boston to Providence appeared as early as 1835, because it cut off the long détour of the sea-way around Cape Cod. It was supplemented in 1837 by a road from Providence to Stonington, Connecticut, which landed passengers and goods at the entrance of the protected course of Long Island Sound and thus avoided the dangerous navigation around Point Judith at the southern extremity of Rhode Island. Otherwise this road, which ran along within two miles of the coast in order to avoid the hundred feet elevation line of the interior, did not greatly shorten the distance between Providence and Stonington. The Sound route from Stonington to New York is safe and direct. Hence we find the railroad from New York to New Haven did not open till 1849, and that from New Haven to New London till 1852. Along the eastern shore of Massachusetts, also, where exceptional maritime conditions had produced a race of seamen, railroads developed somewhat slowly in competition with coastwise communication. The Eastern Railroad, consolidated from several minor lines running directly northward near the coast, was opened between Boston and Portsmouth, New Hampshire, in November, 1840, and extended two years later to Portland, Maine. This road served the manufacturing area of eastern Massachusetts and competed with the longer sea route around Cape Ann; but its branch running from Beverly out to Gloucester and Rockport on the extremity of this peninsula was not constructed till 1847. Plymouth waited for rail connection with Boston till 1846.

South of New York Bay, more devious sea routes between the leading tidewater cities, due to the long projections of the New Jersey, Maryland, Virginia, and Carolina peninsulas, and incidentally the location of the national capital on the medial one of the series, early determined the construction of a chain of railroads across their base. The deep-running estuaries of this stretch of coast necessitated the location of the railroad approximately along the "fall line" from New York south to Goldsboro, North Carolina; but there it left the one hundred foot elevation line and dropped down to Wilmington at the mouth of the Cape Fear River.

Following this road in detail, we find that the earliest link in the chain reflected the active intercourse between New York and Philadelphia. It ran from Camden, on the Delaware River just opposite Philadelphia, 64 miles to South Amboy on Raritan Bay, was begun in 1830 and finished in 1834, four years before the completion of the Delaware and Raritan Canal. In 1839 a second railroad (31 miles) was opened between Bordentown and New Brunswick, the termini of the canal, parallel to the first road in its eastern half. A line opened between Jersey City and New Brunswick (34 miles) in 1838 completed the communication with the Hudson River and New York.

From Philadelphia southward the railroad, following closely the line of one hundred feet elevation, was opened in 1837 through Wilmington, Delaware, to Baltimore; there it connected with the Washington branch of the Baltimore and Ohio, which had already been in operation three years. From Washington the

Potomac, owing to its southward course as far as Acquia
Creek, long persisted as a link in the line of communication between the national capital and Richmond. A road
was opened in 1837 between Richmond and Fredericksburg on the Rappahannock near the bend of the Potomac, but not till 1872 was the line extended to Washington and the steamer trip from Acquia Creek dispensed
with. A writer in 1873 deplores this as the sacrifice of
" one of the most pleasurable features of the trip between
Washington and Richmond." [See map of Virginia
rivers, chapter xiv.]

From Richmond a short line (22 miles) was finished
southward to Petersburg in 1838, to complete connection with an existing road (since 1833) which ran closely
along the one hundred foot elevation line across the base
of the deep peninsula of southern Virginia directly
southward through Emporia to Weldon, North Carolina.
Difficulty of communication down the rivers of North
Carolina, the long détour around Albemarle Sound,
Chesapeake Bay, and the James River had undoubtedly
dictated this cut-off road. The continuation of the line
southward through Goldsboro to Wilmington was not
made till 1840, a fact which reflects the slow economic
development of North Carolina at that time.

To the geographer the interesting fact in this long
chain of railways from New York City to southern
North Carolina is its close adherence to the line of one
hundred foot elevation. This is the line marking cities,
mills, and the head of navigation on the rivers or their
estuaries, retreating farther and farther from the seaboard from the mouth of the Hudson southward. This

western limit of the coastal plain presented few problems
to railway building beyond the bridging of the wide
drainage streams, which in early days were crossed by
ferries from station to station.   The smooth, even slope
of the Piedmont upland and the broad trough of the
Appalachian Valley also offered few obstacles.   Hence
as the development of the hill and interior valley country
progressed, these natural longitudinal routes were traced
by a line of rails from northern Georgia to the Potomac.
The Piedmont road, now the Southern system, runs at
an elevation between five hundred and one thousand
feet from Manassas Junction, whence it connects with
Washington, southwest through Culpeper, Orange,
Charlottesville, Lynchburg, and Danville in Virginia ;
Greensboro, Salisbury, and Charlotte, in North Carolina ; Spartanburg, Greenville, and Westminster, in
South Carolina, to Atlanta, Georgia.

Piedmont railways are a feature of every map.   They
mark the limit set by nature to economic construction
of great routes of transportation along the flanking
plain.   Their elevation varies with the magnitude of the
mountain system which they trace, and the surface features of the plateau base which they traverse.   We see
the piedmont railroad of the Atlantic slope reproduced
in the piedmont line at the eastern foot of the Rockies,
running from Calgary (3388 feet elevation), at the base
of the mountain wall in western Alberta, southeast
through western Montana, where along the sources of the
Missouri River, the Yellowstone, and the Big Horn it
rises at times to an elevation of five thousand feet;
thence south along the Laramie Range to Cheyenne,

and still at the same altitude along the eastern base of the Colorado Rockies through Denver and Pueblo to Las Vegas, whence it turns westward across the mountains by the old Santa Fé Trail to the valley of the Rio Grande, to follow this stream southward to El Paso.

Only the long, gradual upward slope of the Great Plains, which places the foot of the Rockies at a considerable elevation, renders possible a railroad of such length at such an altitude. The steep grade of the Alps and the Himalayas from the Po and Ganges respectively eliminates the piedmont plain as a physical feature; hence the piedmont railways run along at a very low altitude. In the case of northern Italy, the deep reëntrant valleys of the Italian lakes conspire with the slope to force the railroad almost down into the plain of the Po. In northern India the piedmont railway begins at Peshawar at an altitude of only 1165 feet in the northern corner of the Punjab, falls to an altitude of 820 feet at Ludhiana, — though Simla at the terminus of a branch road only about sixty miles northeast of this point lies at an elevation of 7116 feet, — to 700 feet at Rampur and to 255 feet at Gorakpur, where the Ganges begins to encroach still more upon the base of the Himalayas.

But unless a mountain system presents an almost insuperable barrier, as in the case of the Caucasus and the Himalayas, a piedmont railway, owing to the less fertile and productive character of the country which it traverses, unless valuable mines supply the motive for its early development, will in general be antedated by transmontane roads, because the over-mountain country with its different climate and products offers a better field for commerce. This was true in the United States in the case both of the Appalachians and the Rockies. Especially is this the order of development when beyond the mountains lies the sea. The purpose to reach Pacific ports pushed both the American and Canadian overland roads.

The period from 1830 to 1860 saw six railroads constructed across the Appalachians at various points between the Mohawk valley and northern Georgia, where the upland system dwindles away into low hills offering little obstruction to railway building. The roads followed with slight variations the lines of least resistance settled upon by the old trails of Indian, trapper, and western pioneer, and the later routes of proposed canals which were constructed in whole or in part, or were abandoned altogether as impracticable. From New York to the southern boundary of Virginia the lower altitude and dissected character of the Appalachian system afforded the first routes for railways which were economically possible; but from Virginia southward for a stretch of three hundred and fifty miles, the massive wall of the Great Smoky Mountains effectively discouraged the construction of a railway over this barrier till 1882.

The route which saw the only through canal system across the Appalachians saw naturally the first transmontane railroad. From the little seventeen-mile road from Albany to Schenectady as an embryo developed the different stages of westward railroad advance along the line of the Erie Canal. Year by year the links in the chain were added, — Utica and Schenectady in 1836,

Syracuse and Utica in 1839, Auburn and Syracuse in 1838, Rochester and Auburn in 1841, Rochester, Lockport, and Niagara Falls in 1838. In 1843 a traveler could cross the state of New York by rail from Buffalo to Albany, but he was carried by sixteen different companies. In 1851 the Hudson River railroad completed rail connection between the Atlantic seaboard and Lake Erie.

To New York belongs the honor of having constructed also the second railroad across the mountains. The Erie road, taking advantage of the dip in the mountains south of the Catskills, started from Piermont on the lower Hudson, taking seven years to crawl over the Highlands to Port Jervis on the Delaware, whence it pushed rapidly westward along the valley of the Susquehanna and its western tributaries, and over the second watershed to Dunkirk on Lake Erie, in 1851, nowhere reaching an elevation of eighteen hundred feet. Thus in this year New York had two lines from the harbor of the Hudson to the western lake.

The Erie was followed closely by the Pennsylvania, and Baltimore and Ohio railroads, opened respectively in 1854 and 1853. The impracticability of a canal between the Potomac and the Ohio forced the development of the more southern road. It reached the old pioneer station of Cumberland, at the northern bend of the Potomac, as early as 1842; but from that point the discouragement to construction over the rough surface of the Allegheny Plateau and the high altitude to which it had to ascend (2620 feet) retarded its completion to Wheeling and the Ohio till January of 1853. The Pennsylvania line extended westward from Harrisburg,

which since 1838 had had rail connection with Philadelphia, followed the canal route up the Juniata River to
Huntingdon and Holidaysburg, passed thence over the
summit of the mountains to Johnstown and down the
valley of the Conemaugh River to the Allegheny, and
down this stream to Pittsburg. Difficulty of construction
over the main crest of the mountains delayed the road
for two years, so that through connection was not made
till 1854.

With the completion of these four roads across the
mountains, the competition of rail and waterways began,
but at first very slowly. In 1852 the through freight
east on the Erie Canal was twenty-six times that carried
by the Central and Erie lines, but in 1853 it was only
fifteen times greater. In one year more the absolute
decline of tonnage on the canal began to be alarming,
so much was being diverted to its rivals. In Pennsylvania the inadequacy of the canal system as a transit
route over a mountainous country had long been apparent, and its early (1857) purchase by the railroad
company eliminated the last vestige of competition it
might have offered.

In the mean time, the opportunity for a railroad around
the southern extremity of the Appalachians along the
route known to the early Cherokee traders was being
exploited, and a chain of many links from Savannah and
Charleston through Atlanta and Chattanooga, the two
" gate " cities, to river navigation at Nashville was completed in 1854, while another line from Chattanooga
through Stevenson, Alabama, to Memphis opened connection with the Mississippi in 1858. The lack of river

communication along this east and west route, the impracticability of Gallatin's proposed canal, the adaptation of the face of the country to railway building, and the need of access from the southern states of the Mississippi basin to the Atlantic seaboard, all combined to push the construction of this southern line.

Chattanooga had now a central position on a great route east and west, and in its rear lay the Appalachian Valley, affording a natural route of communication with the northeast. Hence we find that the next transmontane line, though starting in Virginia, had Chattanooga for its objective. From its seaboard terminus, Norfolk, at the entrance of Chesapeake Bay, it ran westward along the low watershed between the James and Roanoke rivers to Lynchburg, which it reached in 1854. Thence it passed through the Blue Ridge by the water-gap of the James, turned southward up the Valley of Virginia by the path of the old Wilderness Road to Bristol on the upper Holston, in 1857, and down the valley of East Tennessee to Knoxville and Chattanooga by 1858. Though so late in being constructed, its value as a link in the great through road between New Orleans and the eastern cities was soon appreciated. Subsequently it became a main trunk line from which one branch turned off eastward at Morristown up the French Broad River to Paint Rock, on the western slope of the Great Smoky Mountains, in 1861, and there in 1882 formed a junction with a line which came up from Salisbury on the eastern piedmont railroad and crossed this forbidding range by way of Swannanoa Gap and the upper valley of the French Broad.

Another branch from the Tennessee valley railroad turned westward to Cumberland Gap, and there united in 1890 with another line passing down to central Kentucky; and a third quite recent one turned westward a little farther north at the elbow of the New River in western Virginia, passed down that stream for about thirty miles, and over the watershed to the Tug Fork of the Big Sandy by the old war-trail of the marauding Shawnees, and continued down that river to the Ohio and the railroads traversing its banks.

In the mean time, however, before these branch roads from the Great Valley thoroughfare had multiplied the rail routes over the Appalachians, another road was planning to utilize the pioneer trail from the western sources of the James over the mountains to the Greenbrier branch of the Kanawha. As early as 1857 a road passing along the head streams of the Shenandoah and the James had reached the base of the main Allegheny range at Covington. The extension to navigation at Huntington on the Ohio was undertaken entirely by the state of Virginia, which still cherished its old dream of being a great transit region, a dream which the river paths of the Virginia mountains had awakened in the minds of Washington and Jefferson. The difficulty of construction down the New River cañon and the interruption of the Civil War retarded the construction of the work till 1873.

Three out of the four earliest trans-Appalachian railroads, as we have seen, followed the routes of finished or unfinished canals; and even after reaching the interior basin of the country, the railroads continued to search out the domain of the waterway as if to challenge its competition. The geographical facts are these. The waterways, whether lake or canal, traced the line of least elevation and hence least resistance. The westward trend of the Great Lakes brought the terminus of the water route near to the natural concentrating channel of the Mississippi River, and the western canals united the Lakes with the Ohio. Both lakes and canals, Ohio and Mississippi, were dotted with the largest cities of the western country. Commerce had been started in its course along these waterways. The limitation of canal navigation, due to shallow draft vessels and the short open season in these northern latitudes, soon proved a serious stumbling-block as western resources were developed and commerce grew. The lines of traffic which the canals had created the railroads fell heir to ; the transmontane roads did not begin to come into their full heritage, however, till railways began to spread over the interior basin between the Ohio and the Lakes, and the produce of western farms could be loaded first into freight cars instead of canal-boats. Thus the strategy of transportation patronage lay in the question of reloading. When railroads by their wide expansion had gained this point, by their rapidity and reliability they surpassed their rival. Hence we find the western trunk lines seeking the path of the waterways.

Just as the Ohio was the first western state to follow the example of New York in canal building, so now she was the first to imitate her in the construction of railroads. Here again geographical location on the constricted area between free navigation on the Ohio and the Great Lakes was a potent factor in transportation development. It made a short passway for the steel arteries of commerce as before for water channels. The Mad River railroad, begun in 1832, was opened in 1848 between Sandusky on Lake Erie and Dayton, a city which already had canal communication with the Lake through Toledo. This road in connection with a previously (1846) completed line, the Little Miami between Springfield and Cincinnati, which was also in canal communication with Toledo, formed the first through line from the Ohio to Lake Erie. A second line in operation in 1851 connected Columbus with Cleveland, and a third opened communication between Cleveland and Pittsburg in 1852.

Thus Ohio had three railroads across the state from north to south when the New York trans-Appalachian lines began to push westward along the southern rim of Lake Erie, from Buffalo to Erie, Pennsylvania, in February, 1852, from Erie to Cleveland later in the same year, and from Cleveland to Toledo in January, 1853. This was the last link in the line between New York and Chicago, because the approximation of the heads of Lakes Erie and Michigan had some years before suggested the plan of a railway short cut across the base of the Michigan peninsula. This line, first projected to run from Monroe just north of Toledo over to New Buffalo (a suggestive name) on Lake Michigan opposite Chicago, was pushed half of the distance to Hillsdale in 1843; but when reorganized under a new company was diverted to Chicago and the road from lake to lake opened in 1852 with the eastern terminus at Toledo.

No sooner had the New York roads connection with the lines of the Lakes and the Ohio basin than an enormous increase of their commerce ensued. To this the state of Ohio contributed a large measure. In spite of slack water navigation on its canals, traffic had previously followed its drainage streams, owing chiefly to the different areas of production in the state. About two thirds of the wheat had sought an outlet by the Lakes and the Erie Canal; corn, which was grown largely in the southern part of the state, and all other provisions went out by the Ohio River to New Orleans, whence they were transported to the Atlantic seaboard. Of the exports of beef from Cincinnati in 1851, the year prior to the through railway communication with the East, 97 per cent. went down the river and only 2 per cent. northward to the Lake. Of the Indian corn, the river traffic got 96 per cent. and the Lake 3 per cent.; of the flour, the former got 97 per cent. and the latter 1 per cent. Lard, pork, and bacon sought these outlets in about the same proportion.[5] A small quantity of all these articles supplied the demand in the growing city of Pittsburg. The heavy movement of provisions to the South was due in part to the fact that animals could not be slaughtered till cold weather had set in, when navigation on the canals was closed. Hence the provision trade was readily diverted to the railroads, and with the growing demand in the eastern markets attending increase of population and of industrial development, it soon came to constitute an important part of the through traffic.

When the Pennsylvania road completed its connection

with Pittsburg, its rail communication with Cleveland was already established. The next year after the Baltimore and Ohio came into Wheeling saw a line opened from Bellaire in Ohio, nearly opposite Wheeling, westward to Columbus; and another road, branching off northwestward from this at Newark, established communication with Sandusky and the Shore line in 1856. Thus the region of the Ohio and the Lakes was rapidly drawn into the sphere of influence of the four eastern roads.

But beyond lay the Mississippi, and already the traffic of this great waterway was being tapped by the Chicago and Rock Island railroad, which was completed in 1854. Prior to this time the canals from Lake Michigan had had little influence in directing the Mississippi trade eastward to the Lake route. All the produce along its western tributaries and most of that from its eastern bank went down the river to New Orleans. The current of the Illinois River carried southward most of the products of central Illinois, and only those from the northern part of the state sought the Lake outlet. But from this time the stream of traffic was deflected at right angles toward the eastern markets. Between 1855 and 1856 Chicago became the center of roads radiating to the Mississippi, — to Galena, Alton, Burlington, Quincy, and the mouth of the Ohio. In 1858 the Milwaukee and La Crosse road struck the Mississippi higher up. The next step was to the Missouri, from Hannibal to St. Joseph through northern Missouri, in 1859, and an extension of the Galena line through Iowa to Council Bluffs in 1866.

The intense localization of our industrial and urban development in the northeastern part of the United States, the concentration here of our great commercial ports, and the distribution of our chief foreign markets across the Atlantic yet farther east in Europe, were the main factors in determining the early east-and-west direction of our principal railroads. When the Mississippi valley was finally opened to the eastern lines, railroad building progressed rapidly, finding in its gentle slopes and low watersheds slight obstacles to construction. The bridging of its wide streams was one of the most serious problems. We notice to-day certain large features of railroad distribution in the United States. The most striking of these is the coincidence of the area of intensest urban development with that of the greatest railroad development. Here the network of steel rails is thickest. It tells the story of incoming raw materials and of outgoing finished commodities, of many mouths to feed, bodies to clothe, houses to build, and of many factories to be supplied with fuel. We notice particularly how all the great lines from the Northwest have to turn the corner sharply at Chicago, owing to the deep indenting inlet of Lake Michigan, and thus produce a congestion of railroads which has been the making of that city. We next observe a thinning of this steel network in the agricultural South and the arid West; the occurrence of certain "vacant spots" where the mesh becomes still wider, spots already familiar as areas of sparse population, in the swamp region of southern Florida, in the lumber region of northern Maine and Minnesota, in the barren Ozark

Mountains of Missouri and Arkansas, and in the rough upland of the Cumberland Plateau in West Virginia, Kentucky, and Tennessee. The few railroads which enter the latter region, with only one exception, strike across the axis of the Appalachians, indicating their primary character as transit lines. The western base of this mountain system, owing to the irregular limit of the Allegheny and Cumberland plateaus, has developed no well-defined piedmont railroad such as outlines the eastern foot of the Appalachians.

We notice lastly beyond the eastern margin of the arid belt a marked predominance of east-and-west roads and a sudden decrease, almost a cessation of railroads running north and south. West of this limit we find only three such roads, — the Cordilleran piedmont already mentioned ; a second line running south from Helena and Butte, Montana, diagonally across the axis of the Rocky Mountains and down the head stream of Snake River to Fort Hall, whence along the western base of the Port Neuf and Wahsatch ranges to southwestern Utah it forms the piedmont road of this rim of the Great Basin ; finally the Pacific Valley line running from Puget Sound along the old trails from the Nisqually to the Columbia and from the Willamette to San Francisco Bay, and on up the San Joaquin valley to Mojave in southern California.

Across the belt of the Great Plains many lines strike out boldly for the Rocky Mountains, following sometimes the banks of the parallel drainage streams, more often seeking the solid ground of the watersheds between. Only six pass the mountain barrier and reach the Pacific; but in view of the youth of the country, its arid and mountainous character, and the sparsity of its population, this number speaks eloquently for the power of attraction of two opposite coasts. Its situation between two oceans makes of the United States a great transit land; hence the amazing development of the transcontinental lines. The alternative of this overland transportation from New York to San Francisco, a distance of three thousand miles, is a voyage over four times as long around the Horn. The wide separation of the Baltic and Pacific coasts of Russia, with the choice of a long voyage through arctic or through tropical seas, has forced upon the Muscovite the construction of the great Siberian railroad. A similar situation in Canada called forth its overland road. All these great lines had a political as well as a commercial motive; the political was immediate, however, and the commercial was largely a matter of faith in the future. The Russian and Canadian lines were purely government undertakings; but in the United States political urgency dictated a gigantic system of land grants and subsidies to the transcontinental roads, which, owing to the character of the country which they traversed, were most expensive in construction and along certain stretches could probably never count upon the support of local traffic.

These overland railroads, for a great part of their way, keep to the old California, Oregon, Santa Fé and Gila River trails. The more northern ones follow for shorter distances the footsteps of Lewis and Clark and other early explorers. But the engineer straightens the routes which nature made and avoids détours which were the trader's only paths. The zonal distribution of these transcontinental roads is interesting. We find two northern lines starting from the western extremity of Lake Superior and contained for all their distance between the forty-sixth and forty-eighth parallels. In the stretch from the Red River of the North to Puget Sound eight cross-lines connect them with the Canadian railways of the Northwest Territories and British Columbia, arteries through which so much rich American blood has been poured into the infant giant of the Far North. The main lines are dotted every few hundred miles with a railroad plexus, — Duluth, Fargo, Moorhead, Helena, Spokane, Tacoma, and Seattle, all of them falling near the forty-seventh parallel, and marking the directness of the connection between the eastern and western termini.

The same phenomenon is observed in the other overland routes. The next one to the south takes its course almost along the forty-second parallel from Chicago through Council Bluffs, Omaha, Cheyenne, and Ogden to the northern end of Great Salt Lake; thence it deviates from its adopted path, sending one branch off along the old Oregon Trail to the Columbia and Portland, another southwest by the California Trail to San Francisco Bay. These two Pacific branches from the Salt Lake oasis promise in the near future (1904) to be augmented by a third, which will follow approximately the old Spanish Trail southwest across the Mohave Desert and Cajon Pass to Los Angeles. Thus Salt Lake City will become preëminently the railroad center of the Great Basin. Already it is entered from the east by the fourth trans-Rocky railroad, which, adhering closely to the thirty-ninth parallel, moves straight across the country from St. Louis through Kansas City, Pueblo, and Gunnison to the Green River just below the outlet of Desolation Cañon, and thence turns sharply northward to join the Wahsatch piedmont road at Utah Lake.

The destiny of the Salt Lake oasis as a concentrating point is assured, because for a stretch of almost two hundred miles across northwestern Arizona the impassable cañon of the Colorado River acts as a barrier to deflect overland railroads either northward towards Salt Lake or southward below the outlet of this mighty trench. Hence we find the next overland route leading directly west along the thirty-fifth parallel from Albuquerque, New Mexico, to Mojave, California, where it unites with the Pacific Valley line. This Santa Fé system is the only road across the great western highlands which has to make a distinct bend, either north to Pueblo or south to El Paso, to establish its eastern connections. The desolate expanse of the Llano Estacado of northwestern Texas, flanked on the west by the barren mesa-dotted highlands of eastern New Mexico and on the east by the undeveloped area of the Indian Territory, has not proved an inviting field for railroad enterprise.

Farther south, the Gila River depression, the gap in the mountain wall at El Paso, the grass highlands of central Texas, and direct connection with the earlier roads eastward from the Mississippi valley around the southern end of the Appalachians, have determined the line of the remaining transcontinental route. This also adheres pretty closely to one parallel, the thirty-third, near which it leaves the Mississippi River and passes through Dallas and Fort Worth in Texas; but then it drops southward one degree for the mountain gap at El Paso and the easy route over the Sierra Madre Plateau, followed by Cooke's wagon-trains in 1846. It strikes the thirty-third parallel again on the Gila River, and this line it keeps to southern California, where mountains deflect it north to Los Angeles and bar it from San Diego, its natural Pacific terminus.

The striking facts in these overland railroads are their distribution in northern, central, and southern pairs; the directness of the lines of connection from east to west; the development of an interior concentrating point in the Salt Lake oasis and of two marginal ones on the Pacific. One of these is located near the Canada boundary about Puget Sound, which receives three of the American transcontinental roads and is in close communication with the Canadian Pacific where it issues from the mountains at the mouth of Fraser River; and the other is near the Mexican frontier at Los Angeles, which to its two overland railroads will in the immediate future add a third coming down from central Utah. This last Pacific entrepôt is as yet only in its infancy, but the acquisition of its new link with the East and the improvement of the harbor of San Pedro Bay will do much to stimulate its development. To San Francisco remains always the advantage of its central location and its direct line of communication with the northern Atlantic seaboard.

The Pacific coast is remarkable for the fairly even

distribution of its active ports along all its great length, — Seattle and Tacoma at about 47° 30′ N. L., Portland at 45° 30′, San Francisco at 37° 45′, and San Pedro, the harbor of Los Angeles, at 33° 45′. In this it presents a striking contrast to the Atlantic coast, which has all of its important seaboard points crowded together between Boston (42° 20′ N. L.) and Norfolk (36° 50′ N. L.), a stretch of five and a half degrees of latitude as opposed to nearly fourteen degrees between extreme ports on the Pacific. The explanation of this phenomenon is to be sought in the various geographic influences already discussed, which have determined economic development and railroad distribution in the whole northeastern section of the United States. Until the last few years the active Atlantic ports were contained within even narrower limits; the inclusion of Norfolk represents a recent expansion from Baltimore southward, concomitant with growing commercial activity in the southern states. But the great seaport development which the future holds for the eastern half of our country is to occur along the Gulf coast, whither the interior railroads are already turning to seek an outlet. The discussion of the causes of this belongs to the next chapter.

NOTES TO CHAPTER XVII

1. From Internal Commerce of the United States, Summary of Commerce and Finance, January, 1901.
2. Ibid. p. 1630.
3. Ratzel, Politische Geographie der Vereinigten Staaten, p. 533.
4. Poor's Manual of American Railroads, issues of 1880–1890, for these and subsequent dates and statistics.
5. Poor's Manual of Railroads, 1881, p. xvii.

CHAPTER XVIII

THE UNITED STATES IN RELATION TO THE AMERICAN MEDITERRANEAN

MEASURED in terms of length and configuration of coast-line, which are the determinants of a seaboard base, and gauged by the productiveness and extent of the contiguous hinterland, the United States has a greater claim to strength as a Gulf power than as a Pacific. Its foothold upon the western ocean measures only 1810 miles, and except for a narrow coast strip, excessive aridity limits the productiveness of the country naturally subsidiary thereto. The deep inlet of the Gulf of Mexico creates for the United States a southern seaboard measuring 1852 miles, back of which lies the rich Mississippi basin, its slope emphasizing the leaning of the country towards the Gulf.[1] The Pacific coast, in view of natural disadvantages, has achieved an amazing development; and yet this is only a promise of what is to come. The possibilities of our Gulf development have never up to recent years been recognized or utilized. The narrow circle of the Caribbean lands has formed the natural field of its foreign commerce. Their products too, like those of the Mississippi valley, are agricultural though tropical. Countries where pack-trains of mules and broncos are the prevailing means of transportation do not offer conditions for extensive exchanges. Where the Gulf ports have sought outside markets in Europe, they have suffered from the competition of the commercially developed centers lying between Massachusetts and Chesapeake bays. The southwestward trend of the Atlantic coast from Cape Cod to the extremity of Florida brings Tampa Bay, our most eastern port on the Gulf, in the longitude of Sandusky, and Corpus Christi in that of Fargo, North Dakota; so that wheat from the famous Red River Valley takes a shorter trip to England than do the products of the Texas fields.

The potential advantages of the Gulf coast over the Atlantic seaboard lie at present only in its greater proximity to Mexico, Central America, and the Caribbean coast of South America; but with the opening of the proposed interoceanic canal this advantage will extend to the whole Pacific coast of the two continents. Already with the forward step of the United States into the Greater Antilles since the Spanish War and the surer outlook of piercing the Isthmus, we notice a shifting of our commercial center of gravity towards the south and the promise of a new historical era for the American Mediterranean.

The Gulf-Caribbean basin was the first part of the Western Hemisphere to be actively developed. The wealth of the Spanish possessions in Mexico and Peru made it a thoroughfare for galleons from Vera Cruz or the port on the Isthmus, where they had received the burdens of heavily laden mules bringing gold over the highway from Panama. This old preëminence passed from it, to be revived for a season when the tide which set toward California in 1849 took largely this channel till the opening of the overland railroad; but it will be fully restored only with the construction of the Isthmian Canal. This historical rise, decline, and rise again of great sea basins is no new thing. The ascendency that once belonged to the Baltic has been lost and so far never regained; that of the Persian Gulf promises to be restored when the inlet becomes the terminus of German and Russian railroads. The sleep which descended upon the Mediterranean at the middle of the fifteenth century was broken by the shouts of de Lesseps' workmen on the Suez Canal, but the old maritime energy has not been fully restored to its bordering lands. The Mediterranean and the Persian Gulf regained their importance when they became avenues to the Pacific, and it is this world ocean which is to determine the final significance of the Caribbean Sea and Gulf of Mexico.

The two Mediterraneans are alike in many respects. Both wash the shores of three great land masses with their tideless waters. Both have two basins each, a northwestern and a southeastern; but the Gulf of Mexico manifests a resemblance also to the Black Sea. Like the latter it lies apart from the marine path of interoceanic travel and is such a land-locked sea that Spain, prior to the loss of Mexico, was geographically in a position to block American commerce passing out into the Caribbean Sea, just as Turkey from its bi-continental location can bar Russia from the Ægean. Both Mediterraneans have seen their several basins, like so many segregated land areas, follow different lines of historical development, resulting largely from differences of geographical location. Each sea has an outer or Latin basin. In the Old World Mediterranean the inner basin has been marked above everything else by the constant recurrence of Asiatic influences emanating from the Phœnicians, Syrians, Persians, Saracens, and Turks; in the New World sea, the inner basin gets its stamp from the youngest, freshest, most progressive civilization of the Anglo-Saxon race. Thus these two present the most marked contrast, and are in a large sense representative of their respective hemispheres.

The two Mediterraneans, though interesting as areas of differentiated local development, focus upon themselves the attention of the world at large because by their east-and-west extension and location in or near the belt of the northeast trade-winds, which was so long the pathway of circum-mundane travel, they form links in the chain of oceanic communication around the world. Both being inlets of the same ocean, they extend the reach of the Atlantic from Suez to Panama. But at these extremities a slender belt of sand and a low line of hills block the path of navigation. These narrow necks of land have debarred the Atlantic side of their respective hemispheres from intercourse with the same countries — the rich, historic lands of eastern Asia; and both have pointed to the expensive and dangerous alternative of doubling a great continent by a long voyage. The American isthmus was discovered because the Asiatic one existed; in trying to avoid Suez, the early mariners ran afoul of Darien.

The American Mediterranean, in spite of its more southern location between the tenth and thirtieth parallels, is quite as near to the center of maritime activity in North America between Baltimore and Boston (39° 20′ to 42° 20′ N. L.) as the Straits of Gibraltar (36° N. L.) to the leading ports of Europe between the forty-ninth and fifty-fourth parallels, at Le Havre, Antwerp, Rotterdam, Hamburg, Bremen, Southampton, London, and Liverpool. And though the Atlantic entrance to the Suez route is located so far north, its exit into the further ocean at Aden (12° 45′ N. L.) falls almost as far south as the Pacific end of the proposed Panama Canal (9° N. L.).

Both Mediterraneans therefore seem destined by nature as great transit basins; but whereas in the European body of water one narrow, easily guarded channel secures connection with the main Atlantic, in the American a dozen or more straits of varying breadth and depth afford free access to the commerce of the world. The Gulf of Mexico, almost encircled by the sweep of broad land masses from the extremity of Florida to the tip of the Yucatan peninsula, backed everywhere by a hinterland of abundant resources, has nevertheless only one direct avenue of approach from the Atlantic, Florida Strait; and the control of this is shared by the United States, Cuba, and the British Bahamas just outside the passage. The Yucatan Channel, the natural line of communication between the mouth of the Mississippi and the Isthmus of Panama, connects the Gulf with the Caribbean.

This sea presents a marked contrast to the northern basin. A narrow rim of isthmus lands on the west and of islands on the north and east make it appear like a mirror in a delicate frame. The thickly strewn islands of the Lesser Antilles border it on the east for four hundred miles, affording harbors and passages in plenty for vessels bound for Caribbean ports or for the Isthmus; for this is the focus of navigation lines in the American Mediterranean. On the north the land frame is more continuous. For twelve hundred miles the rather compact line of the Greater Antilles forms the containing wall from the western point of Cuba to the eastern end of Porto Rico, except where broken by the Windward and Mona passages. Hence this northern rim of the Caribbean presents the more highly strategic positions from the military standpoint,[2] and to the United States the more important locations from a commercial, because its channels form the natural avenues of approach for all vessels from New York seeking the Isthmus of Panama.

Europe first gained a foothold in the Western Hemisphere along the Atlantic or island rim of the Caribbean Sea, and there she longest maintained her claim. Until five years ago, this eastern marge belonged to Europe and monarchy, the western to America and democracy. Wedged in between Spanish Cuba and Porto Rico, republican San Domingo presented as much of an anomaly in its colonial surroundings as formerly democratic Switzerland in the heart of feudal Europe. The only exception to be found on the mainland shore is in British Honduras. Of the original Spanish empire in America, Cuba and Porto Rico alone remained to the descendants of the first discoverers; but now they too have slipped from the nerveless grasp of a decadent despotism. North of these lie the Bahamas, which England appropriated when her colonists settled the Carolina coast near by, and which now are island remnants of vast continental possessions reaching to Florida and the Mississippi. To the south is Jamaica, which Cromwell took to adorn the cap of the Commonwealth. In the Lesser Antilles, British, French, Dutch, and Danes have established themselves at the cost of Spain. Formerly even Sweden was represented by the island of St. Bartholomew. The British possessions in this group are most numerous and extensive, though the French hold the two largest and most densely populated islands, Guadeloupe and Martinique.

Islands are detached areas physically and are detachable areas politically. They tend to fall to the nearest political domain; this is what we may call the politico-geographical law of gravity. But the attraction of a larger and stronger country may prove more potent than that of a nearer but smaller land. According to this law Corsica belongs to France, Sardinia and Sicily to Italy, though Sicily in its checkered political career has experienced the sway of Carthaginians, Romans, Saracens, Spaniards, and Italians. The recent Cretan war was in certain aspects an expression of this politico-geographical law of gravity. The American annexation of Hawaii illustrates the same principle. Cyprus acknowledged once the supremacy of Phœnicia, later that of Turkey, which holds the near-by coast of Asia Minor and Syria; but recently it was alienated to Great Britain, which was operating on another geographical principle, the control of strategic points on a transcontinental waterway. Hence the world at large, and

especially the most interested parties, have looked to
see the Greater Antilles obey the law of their geographical location and fall to the United States; and there
are some who anticipate that the same great magnet
will eventually draw to itself the other fragments of
European empires in the Caribbean Sea.

The isolation which makes an island readily detachable, and its prevailing small size, which renders it easily
controlled from within and, with the exception of those
occupying important strategic positions, less an object
of conquest from without because of its limited resources, have combined to maintain European powers
in the West Indies. "Such a parti-colored sample card
of political areas as is afforded by the Antilles is the
expression of the geographical independence of islands;
it makes a contrast to the magnitude of the political
territories which we elsewhere find in America." [3] This
bizarre arrangement would be impossible on any continent to-day. But these island fragments of broken
empires are found everywhere. Tiny St. Pierre and
Miquelon off the southern coast of Newfoundland are
all that remain to France of its vast Canadian domain.
The Channel Isles are the last geographical evidence of
England's former dominion in France, as Cuba and
Porto Rico were of Spain's supremacy in continental
America. First detached from the common history of
Mexico, Central America, and Peru, they are now de-
tached in turn from the destiny of their mother country
and follow the law of their geographical being. Al-
though only Porto Rico has become a part of the
United States, the limited suzerainty established over

Cuba by the terms of the treaty regulating her foreign relations brings the island in effect under the operation of this principle.

The expulsion of Spain from the Antilles is only the finishing touch in an historical process which has been going on now for a century, and which the United States has constantly aided and abetted. The only astonishing thing is that an operation which was so early accomplished on the mainland should have been arrested when extended to the islands. Geographical location and isolation have been two factors in bringing about this result.

West Florida was acquired from Spain by the natural expansion of the American people, just as Texas was from Mexico. But there is a vast difference between proximity and contact. Cuba, although so near the mainland, has felt the isolation of an island environment. The hundred-mile stretch of Florida Strait constituted an absolute boundary. This was sufficient to bar an influx of American immigration like the steady lapping of the human tide across the Mexican border. Isolation robbed Cuba of adequate support from American volunteers in its uprisings. Filibuster expeditions by sea could be more readily checked at the port of departure or arrival than the silent slipping of resolute men across a land frontier several hundred miles long. Moreover, the peninsula of Florida, which seems almost to bridge the sea to Cuba, was in fact almost wholly inoperative as a connecting link because of its extensive swamp lands, which render the lower third of the peninsula almost uninhabitable. There was here, therefore, no chance of an increasing population which should outgrow the narrow limits of a peninsula and overflow into adjacent islands, as exemplified everywhere else by peninsula history. By the everglades of Florida and the barrier of the Strait, Cuba was removed nearly three hundred miles from the American frontier of settlement.

Nevertheless the interest of the United States in Cuba was close. By its size and location it presented a parallel to Ireland's politico-geographical relation to England. Its acquisition by a strong foreign power was always to be dreaded. When in 1823, it seemed that Cuba and Porto Rico were to be the price of England's support of Spain in a war with France, John Quincy Adams thus wrote to our minister at Madrid: " These islands are natural appendages to the North American continent. One of them, Cuba, lying almost within sight of our shores, is an object of transcendent importance to the commercial and political interests of our Union. It commands the entrance to the Gulf of Mexico and the West Indian seas. The character of its population, its situation midway between our southern coast and Santo Domingo, its safe and capacious harbor of Havana fronting hundreds of miles of our coast destitute of such ports, the nature of its productions and its wants furnishing the supplies and needing the returns of a commerce immensely profitable, give to Cuba an importance in our national affairs with which no other foreign country can be compared. Such are the interests of that island and this country — geographical, moral, and political — that, in looking forward to the probable course of events for half a century, it is impossible to resist the conviction that the annexation of Cuba to the United States will be indispensable to the Union itself. . . . The transfer of Cuba to Great Britain would be an event unpropitious to the interests of the United States."

The annexation of Cuba was a constantly recurring proposition from 1807, when the seizure of the Spanish colonies in the Gulf region by either England or France seemed probable, up to 1898. But the United States had committed itself to a continental policy which it was loath to abandon, though statesmen again and again declared for island expansion. Jefferson advocated the acquisition of Cuba, because it could be defended without a navy, but opposed any further seaward advance beyond the mainland.

The influence of the slave power came in to complicate the question, forbidding the liberation of Cuba by Mexico and Colombia in 1825 because the proximity of the island to the southern states made it an undesirable neighbor if a free-soil republic; Hayti was near enough. Later this very proximity, the adaptation of Cuba's products to the plantation system, and hence its ready incorporation into the body politic of the South, became reasons for its acquisition as a slave state in the view of the Southern party. This of course was opposed by the North; but from 1848 till the Civil War, repeated offers to purchase the island and numerous filibuster expeditions for its conquest attested the eagerness of the party in power to annex the island.

After the war, between 1867 and 1869, the advent of a spirit of expansion led to plans for the acquisition of San Domingo and the Danish Isles; but the American people as a whole were apathetic about the advantages to be gained and the old continental policy was not abandoned. Nevertheless, from the time (1868) Cuba began to make a sustained effort to secure her freedom, the United States became deeply concerned for the fate of the island.[4] The Cuban question, which had been almost perennial since 1807, now became a daily one in the discussions of the President and his cabinet. Cuba lay just at our doors, and everything which influenced it fundamentally touched our commercial interests, which had been growing steadily since 1850. The peace and prosperity of this market of purchase and sale were of moment to our country. Moreover, the proximity of our shores made the United States a natural base for filibustering expeditions, which, in spite of deeply aroused sympathies, the government was bound to prevent. Our ports became the refuge for Cuban exiles, the home of the Cuban Junta. Persecution started an exodus of Cuban subjects to the United States, where they became naturalized. Consider the large proportion of these people in the southern counties of Florida. From their adopted home they returned to transact business in Cuba, but their naturalization papers did not always outweigh their Spanish names and swarthy complexions to save them from mistreatment at the hands of Spanish officials. Thus the close relations, ethnic, commercial, and political, which grew out of geographic conditions, had their logical consequence in the expulsion of Spain from her last foothold in the Western Hemisphere.

From the war with Spain the United States issued as a Caribbean power. Hitherto limited to the segregated basin of the Gulf of Mexico, by her virtual protectorate over Cuba, the pledge of naval stations on its coast, the possession of Porto Rico and the two small islands in the Virgin Passage, she now has a base of twelve hundred miles along the northern rim of the Caribbean. This base will doubtless be extended in the not remote future by the purchase of the Danish Isles. The possession of St. Thomas will give us control of the Anegada Passage, the doorway for most of the steamship lines from Europe and the last important channel into this sea from the north.

The American advance into the West Indian sea derives its great significance from the paramount interest of the United States in an interoceanic canal which shall connect the Atlantic and Pacific shores of the country and secure closer communication with our new island possessions in the Orient. The war which gave us Porto Rico, and thereby raised our geographical status in the American Mediterranean, created also a more urgent necessity for a perfected transit route to the Pacific, magnified our interest in the canal, and at the same time increased our power to control it.

The configuration of that part of the North American continent which forms the western rim of the Caribbean Sea complicates the question of the location and construction of a transit channel. Nature unequivocally limited the Suez Canal to its present site; but geographical conditions in this western Mediterranean are quite different. The American isthmus is about fourteen hundred miles long, extending from the Isthmus of Tehuantepec in Mexico to the Atrato River in western Colombia. For the southeastern stretch of six hundred miles the width of this isthmus is comparatively narrow, varying from a minimum of barely thirty miles to a maximum of one hundred and twenty miles, contracting to this last width again farther north at the Bay of Honduras and at Tehuantepec.

The outlines of the map therefore would indicate numerous possible canal routes. A broad ridge of seven hundred feet elevation, too high and wide to be cut, eliminates Tehuantepec, though its nearness to the United States speaks eloquently in its favor. This was the line of Eads's proposed ship railway. A mountain range disposes of the Bay of Honduras. In southern Nicaragua, the San Juan River and Lake Nicaragua, forming links in an interoceanic route, and a passway over the mountain divide at only 153 feet elevation promise better conditions; they involve, however, a rather long (184 miles) transit route and the construction of artificial harbors at the termini. Three possible lines, known as the Panama, San Blas, and Caledonian, terminate in the Gulf of Panama on the Pacific side. The first, with a summit elevation of less than three hundred feet and a width of thirty-six miles, though necessitating a canal of forty-nine miles, seems to have most in its favor. The last two have the advantage in shorter length and superior ports on the Atlantic; but the height of the dividing ridge (681 ft.) would involve a tunnel canal which would outweigh the other advantages. Still farther south, the Atrato River, which is only seventy miles from tidewater on the Pacific at the nearest point, offers a possibility.[5]

In the early days of Spanish supremacy the long Atrato valley was a natural transit route especially between the Gulf of Darien and Peru, but Philip II. issued one of his genial decrees forbidding the navigation of the river on pain of death. The San Juan River was a regular line of communication for Spanish vessels plying between Cuba and Granada on Lake Nicaragua up to 1639; and again in 1850, when the rush of the gold-seekers to California put a strain on Isthmian transportation, this route was brought into requisition. Likewise the Chagres River, which is navigable for skiffs and light-draft vessels halfway across the Panama Isthmus to Cruces, was opened in 1534 for the Spanish trade, served later for California emigrants, guided the surveyors of the Panama Railroad, and later the engineers of the Panama Canal. Thus history repeats itself along even an insignificant watercourse which serves an important connection.

Owing to the threefold predominance of water on the earth s surface and the consequent development of marine transportation as commerce tends more and more to encircle the world, an isthmus soon loses its importance as a link between land masses, and becomes instead a barrier between water areas, which political and commercial exigencies require to be converted into an artificial strait. The Isthmus of Panama served as a passway for the aimless wanderings of the primitive tribes of Central and South America and was a factor in early ethnic distribution. It may carry the road-bed of the intercontinental railroad which some day in the remote future is to connect the United States and the seaboard of Argentina. But in all this interval of centuries it has acted only as a barrier to interoceanic navigation.

The supreme interest of the United States in an Isthmian Canal, the strength of her geographic location in relation to the same, her natural office as guarantor of the neutrality of the channel and the political stability of the country through which it passes, finally the abundant wealth, the resources of the Anglo-Saxon republic, and the steadily developed aptitude of its citizens for vast enterprises, all combine to lay the task of its construction upon the United States. The geographic influences which brought forth the fishermen of Gloucester, the spinners of Rhode Island, the fur-trading expansionists of the Far West, and the courageous manufacturers of the Pacific states have been only so many remote antecedent causes, stable, persistent, of the American control of the Isthmian project to-day.

This control involves naval supremacy in the American Mediterranean. No other country has here a base so broad and full of natural resources. The long sweep of mainland and islands from the Rio Grande to St. Thomas leaves little to be desired, except the possession of the Danish Isles. The recent extension of the American coast-line following the acquisition of Porto Rico and the Philippines, the rapid expansion of American foreign commerce, and the creation of a trans-Pacific colonial trade, finally the construction of the Isthmian

Canal, which will constitute in effect a bit of American seaboard, must call forth a navy whose strength shall be commensurate with the importance of our extra-continental interests. Such a navy will insure our command of the Caribbean Sea.

Great Britain's holdings in these waters are numerous and well distributed for naval control. Especially Jamaica, by its almost central location in the Caribbean, occupies a strategic position similar to that of Malta in guarding the approaches to the Suez Canal. The limited resources of England's West Indian possessions are compensated for by the political and maritime ascendency of the British Empire. France's positions are all on the outskirts. So are those of Denmark, though the pivotal location of St. Thomas at the angle of the Greater and Lesser Antilles lends it no small significance.[6]

The position of the Dutch, also, near this northeast angle in St. Martin (half French), St. Eustatius, and Saba Isle, and in Buen Ayre, Curaçao and Oruba off the northwest coast of Venezuela, gives them a near and remoter base in the approaches to the Isthmus of Panama. This excellent distribution reflects the interest of Holland in an alternative line of communication with her vast colonial possessions (730,000 square miles) in the East Indies, should the Suez route be at any time impassable for her. The holdings of the French and British in the Caribbean Sea also gain an added significance when viewed in relation to the territories of these nations in the Orient. Only the Danish Isles lack this character of way-stations and hence could be

transferred to some other country without loss of pres-
tige to Denmark. Now Germany, the rising young
colonial power of Oceanica, is accredited with designs
upon a coaling-station either in St. Thomas or Curaçao,
a rumor which gains color also from the astonishing
strides of German trade in Central America and the
exploitation of this field by the German merchant
marine.

While the United States has abundantly proven her
political ascendency in the American Mediterranean, and
while her naval strength there has been sufficient in the
past and will undoubtedly be made adequate to the needs
of the future, she does not show the commercial development in this region which her proximity and facility of
communication, the adaptation of her exports to the
needs of its population, and her extensive importation of
its tropical products, would lead one to expect. The balance of trade with the countries and islands bordering
on the Caribbean is almost always against the United
States. The proximity, however, has been in the past
perhaps more apparent than real, in consequence of the
concentration of the industrial and commercial activities
of this country in the northeastern states. The distance
by sea from New York to Vera Cruz, Mexico, or to
Greytown, Nicaragua, is over two thousand miles. New
Orleans is little over half as far from Greytown and only
788 miles from Vera Cruz; but the export trade of our
Gulf seaboard has only recently begun to be an important factor in our foreign commerce. Canada, which,
besides speaking the same language as ourselves, is contiguous to our area of greatest development in point of

industries and commerce, railroad, lake, and marine transportation, is the best customer that the United States has except Great Britain and Germany. In 1901 the United States furnished 61 per cent. of Canada's total imports for consumption.[7]

In the countries south of our border, most of them speaking other languages than our own, the record of all but the recent past is far from satisfactory, though the last few years have marked the beginning of a change. But in general the growth has been slow and the rate has been retarded with every increase of distance from our frontiers. In contrast to our active trade with Canada, Mexico, which is equally near and enjoys railroad as well as ship communication, has purchased in general only about 40 per cent. of her imports from the United States. However, this proportion rose to 50.6 per cent. in 1900, and to 54.3 per cent. in 1901. In the last year only the American and German trade showed a decided increase, while almost every other country showed a decline.[8]

One change for the better in American business methods in Mexico is undoubtedly due to our recent acquisition of former Spanish possessions. Prior to the Cuban War we were distinctly a one-language people. Our geographical isolation has operated against our becoming linguists. Marked ethnic differences have prevented any but limited affiliation with the nations south of our border, and they are our only neighbors speaking a different language except the negligible quantity of the French *habitant*. Now we have a large number of Spanish-speaking citizens, and the importance of Spanish is growing in our consciousness. Fully half of the American trade circulars now sent to Mexico are written in that language, and many business houses are employing young Mexicans or Cubans as traveling salesmen, with excellent results.

The Central American states, which are a little more remote from us than Mexico, though readily accessible by water, have gotten only about 35 per cent. of their imports from us; in the year 1900, however, the proportion showed a pronounced rise in the more important states. We supply 33 per cent. of her imports to Colombia, 27 per cent. to Venezuela; and to the West Indies, whose trade is naturally reserved for the various mother countries, only about 20 per cent.[9]

In South America, especially that portion lying south of the Equator, our commercial situation is yet more discouraging. The bold eastern projection of the continent which brings Cape St. Roque some forty degrees of longitude, or twenty-six hundred miles east of New York, has the effect of placing New York, Hamburg, and Liverpool at about equal distances from South America and therefore so far on an equal footing of competition in the markets south of Pernambuco on the Atlantic side and all along the Pacific coast. But almost all the steamship lines to South American ports are controlled by European capital. They carry European merchandise to the southern markets, exchange their cargoes for rubber, coffee, wool, and hides, which they bring up to our Atlantic seaboard cities, reload there again with grain and provisions, since the United States exports to Europe so much exceed the imports from that continent, and having reached the home ports, after making the vast circuit (Plymouth, Montevideo, New York, Plymouth — fifteen thousand miles), take on a fresh cargo for South America again. By this system, New York's goods shipped to Montevideo by way of Europe are at a disadvantage, in competition with English or German merchandise having direct steamship connection with the port of destination. The obvious remedy for this condition is to build up a native merchant marine. In all Gulf and Caribbean countries with which the United States now have adequate steamship communication, they have won a reasonable share of the commerce.

The construction of the Isthmian Canal, whereby an almost due south route will be opened from New York and New Orleans to Callao (the port of Lima) and Valparaiso, will greatly favor United States trade on the Pacific coast of South America over that of England and Germany ; and by establishing communication with the trans-Andean railroad, in which only one short link of forty-six miles remains to be completed, it will bring American products into the rapidly developing temperate belt between Valparaiso and Buenos Ayres, and make them compete with European goods in the better populated districts of Argentina, Paraguay, Uraguay, and southern Brazil.

The steady industrial and commercial growth which has been going on in the southern states in the past decade has been suddenly accelerated by our increasing interests in the Caribbean. Already our close relation with Cuba and Porto Rico is tipping the balance of commercial activity more towards the Gulf seaboard. The evidence is to be had in the statistics of exports and imports for New Orleans, Galveston, Corpus Christi, Mobile, Pensacola, Tampa, and Key West.[10] New Orleans, as the focus of converging trade lines in the Gulf and the natural outlet of the Mississippi basin, promises to become one of the great exporting points of the world. In 1901 it supplanted Philadelphia in the third place in the list of grain-exporting cities, being outranked only by New York and Baltimore; and in the fiscal year ending 1902, its grain exports were 36 per cent. higher than those of Philadelphia and very nearly equal to those of Baltimore.[11] The last two years have seen the total foreign commerce of the Crescent City increase 56 per cent.

The rapidly growing importance of the Gulf seaboard is further attested by the eagerness of certain big railroad systems of the Mississippi valley to reach New Orleans. Their purpose reflects the existing trend of exports away from the Atlantic seaboard cities and towards the Gulf; and the consummation of projected railroad combinations centering about Memphis and Fort Worth will start a new current of grain away from Chicago and the East to the Mississippi port. A suggestion, made recently by the president of a great northwestern railroad system having its southern terminus at St. Louis, that the United States government improve the Mississippi to secure twelve-foot navigation from New Orleans to the mouth of the Missouri, is another straw showing the drift of the commercial current towards the Gulf. There is now every promise that the southern seaboard will come into its own, and that the geographical location of the United States in the American Mediterranean will be exploited to the full limit of its possibilities.

NOTES TO CHAPTER XVIII

1. Ratzel, Politische Geographie der Vereinigten Staaten, p. 17.
2. Mahan, The Interest of America in Sea Power, p. 303.
3. Ratzel, Anthropogeographie, p. 575.   1882.
4. For relations of the United States to Cuba, see Albert Bushnell Hart, Foundations of American Foreign Policy, pp. 108–133. 1901.
5. Report of the Isthmian Canal Commission, 1899 to 1901, pp. 69–72. Washington, 1901.
6. For strategic aspects, see Mahan, Interest of America in Sea Power, chapter viii.   1897.
7. Review of the World's Commerce for 1901, p. 129. Washington, 1902.
8. Ibid. p. 131.
9. American Commerce in Mexico, Central and South America, p. 505. Summary of Commerce and Finance, August, 1901.   Washington.
10. Statistical Abstract of the United States for 1901, passim.   Washington, 1902.
11. Bulletin No. 12, Series 1901–1902, Export of Domestic Breadstuffs, etc.   Bureau of Statistics, Washington, 1902.

CHAPTER XIX

THE UNITED STATES AS A PACIFIC OCEAN POWER

HISTORIC areas of civilization are most naturally indicated by the seas which they encompass. The Ægean, Mediterranean, the Atlantic, Pacific mark eras of progress, of which the seafaring nations of the world have been the first apostles. Every maritime expansion of civilization draws with it in its train the smaller water area which constituted its previous field, operates for a long time under the influence of the latter, and while robbing it of its former preëminence, contributes finally to its absolute activity. This has been the history of the Ægean in relation to the Mediterranean, of the Mediterranean and Baltic-North in relation to the Atlantic. We speak of the Pacific as the " ocean of the future," but this means in reality the final expansion of the maritime field by which it embraces this last great basin and advances to active exploitation of the world ocean. For the sea is always one. The little harbor of Batum at the innermost corner of the Euxine is just as much a port of the world ocean as New York, Honolulu, or Hongkong.

When the commercial and maritime drama of Europa was shifted from the stage of the Mediterranean to the Atlantic, those nations who had the front seats got most out of it, as we have seen. They furnished the best trained actors in the stirring scenes of exploration, colonization, and trade, and drew in the largest rewards. Their advantages were fundamentally the product of geographical location. The same principle must hold on the "ocean of the future;" but the development of the world ocean will mean the exploitation of the Pacific from the basis of the Atlantic. The preëminence which the Atlantic has gained will long dominate the Pacific, and geographic conditions make it doubtful whether this supremacy will ever pass to the larger basin. Therefore, those countries which have a foothold on both these oceans possess the vantage-ground; and their potential strength will be in proportion to the length and proximity of their two ocean frontages and the resourcefulness of their respective hinterlands.

A narrow ocean, near-lying continents, remote watersheds, long navigable river systems, accessible inland regions, a large back country to draw upon — that is the Atlantic field. A vast ocean, remote continents, a few, fall-broken rivers, mountain walls hugging the coast, an inaccessible interior, limited back country — that is the Pacific field. The Pacific, though twice the size of the Atlantic, has a drainage basin less than half as large. The reason for this disparity lies in the fact that the primary highlands of Asia, Australia, North and South America are situated on the Pacific coasts of those continents. In Australia and South America the mountains rise directly from the sea, and only short, plunging torrents erode their slopes, while all the extensive drainage is in another direction. In the northern continents the watershed is a thousand miles or more inland, and thereby furnishes almost the whole drainage area of the Pacific; but this location of the divides does not mean navigable streams to the coast. The unfavorable character of our American Pacific rivers has been explained. Those of Canada are no better. Only the far-away Yukon, by sweeping around to the north of the coast range, affords a navigable course for steamers of light draft for 1370 miles to Forty Mile Creek, where the international boundary line of Canada and Alaska crosses the river; but the nature of the country through which the Yukon flows and its ice-bound condition most of the year render it of little commercial importance except to the Klondike miners.

The Asiatic rivers of the Pacific are much more important. The Amur with its tributaries affords hundreds of miles of navigable waterways, but the sharp northward bend of its course just before reaching the ocean discharges the stream into the Okotsk Sea, where its port is frozen six months of the year. The next great river, the Hwang-ho, though flowing with a sluggish current through the great plain of eastern China, is too shifting in its course to be relied upon for navigation. The Yangtse-kiang alone is comparable to the great streams of the Atlantic. It is navigable for one thousand miles from its mouth and admits even ocean-going vessels six hundred and thirty miles up to Han-Kau, where they take on cargoes of tea and silk for Europe and America. In the Si-kiang and Mekong navigation is much impeded by rapids. Therefore of all the rivers flowing east and west into the Pacific, only the Yangtse-kiang affords a good route of communication

between seaboard and interior. Hence we find it lined
with free ports all the way from Shanghai, on the coast,
and Chin-kiang, at the head of its estuary, to Ichang,
a thousand miles up its course.[1] The Yangtse-kiang
is the one valuable river adjunct of sea power in the
Orient; hence the discerning eye of the English early
appropriated it as the British "sphere of influence."

China, by reason of its long irregular coast-line stretching through twenty-one degrees of latitude, the possession of the one navigable river of the whole Pacific, its
central geographical location in the temperate zone, and
its large territory of abundant resources, has a strong
position on the Pacific; but dominated by a nomad
people from the inland steppes of Asia, bred to isolation by the great sweep of mountains behind them, and
unreceptive to the vitalizing influences of the Atlantic civilization, the Chinese have not profited by the
advantages of their location. Russia's frontage on the
Pacific finds its value much reduced by its subarctic
situation. Japan has central location, a long island
base, and the spirit of progress which is developing at
a wonderful rate the maritime activity of the kingdom;
but it lacks area and population, important factors in
political and commercial strength. England's possessions in India, Australia, New Zealand, North Borneo,
New Guinea, Hongkong, and Malacca give her a broad
enough base in the Pacific; but their scattered location, remoteness from the national centre of strength
in the Atlantic and also from the storm-center of the
"problem of Asia," between the thirtieth and fortieth
parallels, all combine to reduce the value of her Pacific

position, though the maritime strength of Great Britain makes her at present the greatest factor in the Eastern Question.

China is the only power on the western shores of the Pacific to whom geographic conditions might have given political and commercial preëminence, except for the one thing needful, contact with the Atlantic. This the United States has, besides the requisites of a long Pacific coast-line, central location in the temperate zone, and a large territory of abundant resources. To balance the teeming plains of eastern China, it has the long Pacific Valley with its enterprising if not dense population. A coast-line of 1810 miles, which rises to 8900 miles if all the inlets, bays, estuaries, and islands are included, a few evenly distributed ports for larger vessels and numerous safe roadsteads for smaller craft, yield a fair degree of contact with the sea from San Juan Strait to the excellent harbor of San Diego, in spite of the mountain range which faces the coast.

The great advantage of the United States is its interoceanic location. This it shares with Mexico, the Central American republics, Colombia, and nominally with Chile, which has stretched a narrow tape of territory around the southern extremity of the continent to the Atlantic entrance of the Straits of Magellan; but obvious geographic limitations of climatic situation, natural features, and size restrict the political and commercial ambitions of all these countries, even if we ignore the inferiority of their Latin-American populations. To the north, a neighbor of the same Anglo-Saxon blood as ourselves has also a broad frontage on both oceans, and as part of the British Empire has also like ourselves mid-sea islands in the Pacific to serve as way-stations to the opposite coasts. British Columbia, with its one thousand miles of seaboard and its excellent harbors, occupies a fine position in relation to trade with China and Japan, Russia and Manchuria. It has also considerable natural resources, especially coal of good quality which rises in importance because of its scarcity along the United States littoral. British Columbia has moreover the resources of all Canada at its back; but when full weight is given to all these favorable conditions,[2] the disadvantage of a location too far north, and hence the restriction of area adapted for the support of a dense population and the production of superfluous wealth which may enter the markets of the world as capital, place British Columbia and Canada far behind the United States in the rivalry for the commerce of the Pacific.

The first weak foothold of the young Republic in the far-away station of Astoria at the mouth of the Columbia River, and that early trade in furs with China, seemed prophetic of the destiny of the nation. The never abandoned purpose to widen the frontage on the western ocean, the obstinate debate of the " Oregon Question," and the conquest of California, committed the United States to the career of a Pacific power. Wide though the ocean is — ten thousand miles at the Equator, eighty-five hundred at the Tropic of Cancer between Hongkong and Mazatlan on the Mexican coast, and forty-seven hundred and fifty between Yokohama and San Francisco approximately along the thirty-seventh parallel — and sparse though the islands are for a belt of two thousand miles off its American shores, the mere presence of the United States on the Pacific has been a sufficient reason for concern in all matters pertaining to this ocean.

This was the ground on which Russia in 1867 pressed Alaska upon us, strengthening our base on the Pacific and weakening that of her hereditary enemy in British Columbia by placing this English colony between the fires of American enterprise on both its northern and southern borders. The " ten marine leagues," moreover, which fix the width of the long " panhandle " of southern Alaska, cut off a thousand miles of the natural Pacific frontage of British Columbia ; while the possession of the peninsula and the Aleutian Islands gives our Pacific base a reach of over four thousand miles from San Diego to Attu, three hundred miles beyond the international date-line of the one hundred and eightieth meridian, and only six hundred miles from the nearest Japanese islands. The acquisition of Alaska gave us Russia for a near neighbor in Bering Strait. The ownership of the Pribilof Islands in Bering Sea, which the seals have adopted as their breeding-grounds, made us the most interested party in the controversy of the seal fisheries, and involved us in negotiations with Russia, Japan, and England, which resulted in the Paris Commission of 1895.

The vast importance of our Alaskan territory is developed, however, only with our advance as an acknowledged world power in the domain of the Pacific by the acquisition of the Philippines. The cause lies in our

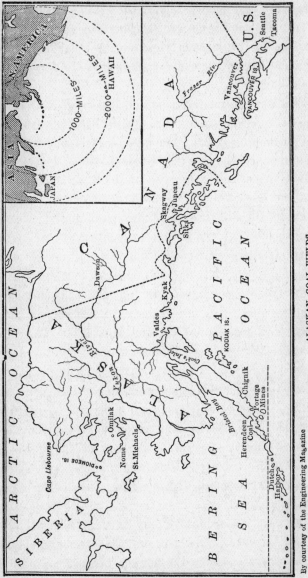

By courtesy of the Engineering Magazine

ALASKAN COAL-FIELDS

increasing need of coal. This Alaska can furnish in
abundance and of an excellent quality. There are out-
croppings of various grades in the islands of the " panhandle," on the mainland at White Horse Pass near
the railroad from Skagway to the Yukon trail, and on
the seaward slope of the Mount St. Elias Range ; but
the first field with promises of real commercial importance is found farther north along the coast, just east
of the mouth of Copper River and behind the island of
Kayak. This coal is the best found on the whole Pacific
seaboard, and is almost equal to the standard Albion
Cardiff coal of Wales. The seams are thick and extensive, but the lack of a protected harbor near the mines
puts a serious obstacle in the way of the English company now developing them. Coal is found also under
commercial conditions at Kachekmak Bay off Cook Inlet
at the extremity of the Kenai Peninsula ; but the most
extensive and important coal-field in Alaska lies still
farther west towards the extremity of the long slender
Alaska Peninsula and on Unga Island, cropping out in
numerous seams at Chignik and Portage bays on the
Pacific, and Herendeen Bay on Bering Sea. Herendeen
and Portage inlets almost nip the peninsula in two, leaving only a five-mile neck of land between them, so that
these two locations are virtually one. The coal ranks
next in quality to that of Kayak, and is equal to any
mined farther south. Because of the situation of this
field near protected harbors, an important matter in the
wide sweep of the Pacific, and its location on the great
circle of navigation constituting the shortest route for
steamer lines between the United States and any part of

Asia, it assumes paramount importance. Portage Bay will be a coaling-station three degrees of longitude farther west than Honolulu and one equipped with its own mines, from which coal will be delivered to deep-sea vessels at low cost.[3] Situated in the latitude of Glasgow and exposed like it to mild oceanic winds, these Portage Bay mines have a winter not more severe than that of Philadelphia ; so climate will interpose no obstacle to their development.

The possession of the Philippines, which lends a new importance to our neglected province of the north, was the signal for us to profit by certain other advantages which lay at our door or within our hands and had been ignored. Dewey's victory in Manila Bay made imperative the annexation of Hawaii, Wake Island, and the retention of Guam, the southernmost isle of the Ladrones group, so that a direct mid-ocean line of communication between the home shore and our oriental colony might be assured. Just as England on the route between London and India acquired first the remote extremity in the Orient, and then filled in the intermediate points — Gibraltar, Malta, Cyprus, Suez, and Aden — which were to secure her long line of communication, so the United States had to turn her attention to getting way-stations across the Pacific. Far back, between 1841 and 1867, she claimed Wake, Christmas, and Midway islands by right of discovery, but entered into active possession of the first only recently, when its geographical location between Guam and Hawaii indicated its natural province as a Pacific cable-station.

Since 1889 the position of the United States as a

Pacific power had been attested by its share in the tripartite government of the Samoan Islands. This share was converted into the absolute ownership of Tutuila and Manua in 1899 by the acquiescence of England and Germany. The fact that the superior harbor of Pango-pango on Tutuila was ceded to the United States as a coaling-station as early as 1872, but was not actively occupied till 1898,[4] bears witness to the American change of policy. The whole Samoan group is of great strategic value. It is situated at about 14° south latitude and 170° west longitude, on the direct path from Puget Sound to Sydney, Australia, and on a line from the Isthmus of Panama to east Australian ports. Herein lies its significance for the United States.

The island world which, like a vast continental nebula, occupies the expanse of the Pacific, becomes more and more rarified in the direction of the American mainland, till it finally leaves a wide gap of ocean space between the outlying groups of Hawaii and the Paumota Archipelago on the two Tropics and the mountain-bound coasts of the Western Hemisphere. Neither North nor South America finds at its western door natural stepping-stones for advance into the Pacific. Hence the Hawaiian group, the only place in the whole Pacific north of the Equator and east of the continental islands festooning the coast of Asia where the nebula has condensed into a constellation, by their location, their magnitude, and their isolation become the paramount strategic position west of the American shores. They afford the first ocean-station on a trans-Pacific line of communication.

Political gravitation has drawn the Hawaiian Islands to the dominion of the United States. Their location in relation to the American shores at an early date made them a place of call for New England trading-vessels on their way to the northwest coast or to China. Their situation at the great cross-roads of the Pacific tended to make their population a precipitate of all the races traversing this ocean, but the American element predominated. An island environment suffers always from the limitation of its size, especially in commerce. England had to go out in search of the markets of the world. Where she found none, she created them by forming colonies. Japan is on the same keen hunt to-day. Cuba cannot survive without the markets of the United States. The nearest market is the logical one because of economy in transportation. Hawaii from the beginning of its sugar culture needed American buyers. In 1851 a strong effort, emanating from the Islands themselves, was made for annexation; but the United States Senate refused to ratify the treaty offered by the Hawaiian Parliament, and instead promised protection. Meanwhile trade between the two countries was growing and with it the sentiment of annexation in Hawaii, while the big Republic held to its continental policy. Sugar became the chief interest of the Islands; it was almost wholly in the hands of Americans, who formed the monied and the ruling class, but who found a naturally remunerative industry almost hopelessly crippled by the heavy duties on their products at the port of San Francisco. Annexation would make them rich, so the agitation was kept up. Here again that extra-territorial expansion which had fixed the destiny of West Florida and Texas was doing its work. In 1876 a compromise was effected. The chief Hawaiian products were admitted to the United States free of duty, and the United States was given a naval station in the vicinity of Honolulu. The barrier was beginning to fall. The Islands were commercially within the sphere of the United States, and the United States had advanced to the strategic outpost on the Hawaiian shores. This status of overlapping boundary was maintained until 1898.

In the mean time, however, this country was not so indifferent to the fate of Hawaii as it seemed. Twice it interfered in Hawaii's behalf with England, and in 1850 it warned off the French. In 1843, when there was danger of acquisition by England, Daniel Webster declared that no other power ought to get possession of these islands either by conquest, or for the sake of colonization. There was much of the dog in the manger in this attitude. Prior to our acquisition of the Philippines, Hawaii possessed far greater political utility for British Columbia than for us, because it was the one East Pacific station on the long intercontinental sea route between Vancouver and Sydney. From a military standpoint, however, it was quite as important to us; because Hawaii as a coaling and naval station, two thousand and eighty miles from San Francisco and not more than twenty-five hundred from any point on our western coast, in the hands of any foreign power would have been a standing threat.[5] So held, it would have paralleled the Bermudas, which were the source of discomfort enough to us in the Revolution and the War of 1812. Hawaii in our hands, every one else is necessarily excluded from the field, except where the British are entrenched in their port at Esquimault, near Victoria on Vancouver Island.

The Hawaiian Islands, Wake Island, and Guam form to Manila a line of communication lying between the narrow limits of the thirteenth and twenty-first parallels. The American termini of this line are located at San Francisco (38° N. L.), Los Angeles (34° N. L.), and Panama (9° N. L.), to all three of which Honolulu holds a central position. The preëminence which it now enjoys as the focus of the great commercial routes of the Pacific will be only enhanced with the opening of the Isthmian Canal, because it will lie in the path of an increasing file of vessels moving along from Panama to China, Japan, or Asiatic Russia. At the western end of this island chain of communications are the Philippines, by reason of which the United States has now become an Asiatic power. This large group, scattered over an area measuring one thousand miles from north to south and half as far from east to west, is located wholly within the tropics, and distributed around it in a wide-sweeping semicircle are the oriental countries whose vast populations make the markets of the East.

To these markets the United States supplies about ten per cent. of the imports and the commercial countries of Europe fifty per cent. Asia and Oceanica are increasing their imports, and a fair share of the increment falls to American merchants. Especially is this true in Japan, where the American trade is advancing by leaps and bounds, and now ranks closely behind that of England. Exports from the United States to Asia have grown from $11,000,000 in 1870 to $64,000,000 in 1902; those to Oceanica have increased from $4,335,000 in 1870 to $34,000,000 in 1902, and the last figure does not include some $19,000,000 worth of merchandise sent to the Hawaiian Islands but not included in this report since their annexation.[6] The total commerce of the United States with Asia and Oceanica rose from $138,000,000 in 1892 to $173,000,000 in 1897 and to $287,000,000 in 1902.[7]

These commercial achievements of the past thirty years have been carrying the United States inevitably forward into the maelstrom of Asiatic affairs, because there lay the center of our new interests. The proximity of Cuba to our shores was the far-away determinant which placed the stars and stripes in the Philippines and accelerated the progress of history; but this or some other base, perhaps on the Chinese mainland, would have been our destiny. The commercial strength of the American Republic was bound sooner or later to find a political expression in that international struggle for existence, which is a struggle for space, going on in Asiatic territory. A chain of historical events, largely geographical in their causes, determined that the Philippines should be the channel of American influence in the East. The detachable character, inner weakness, and protected isolation of every island group maintained Spain in her insular possessions here, as in the Antilles; so that the blow which despoiled her of one took all.

Now this same protection against international entanglements which is yielded by an insular position accrues to the United States, makes her situation in the Orient analogous to her continental location at home, and gives her the best possible base from which to protect her interests.

These interests, chiefly of a commercial nature, are well worth guarding. The needs of the Orient are those of an old, crowded country which, however, has not advanced into the modern industrial stage of development. The demand comes therefore for grain and flour, refined oil, wood and leather manufactures, tobacco, cotton goods of all kinds, agricultural implements, and all forms of iron and steel manufactures. These the United States can provide; and the advantages arising from abundant natural resources, the extensive organization of industry which reduces the cost of production, and the superior character of our industrial methods, enable the American merchant to compete in these distant markets with other nations who operate from a base thousands of miles nearer. The vast size of the home country, affording a wide field for raw materials, for concentration, and hence organization of industries, for the development of transportation facilities, and the maintenance of an extensive and varied domestic market, has been in the past the dominant factor in the evolution of the American entrepreneur. The other factor has been his Anglo-Saxon vigor of character and tenacity of purpose. Here heredity and environment have combined to do their utmost, and the result has not been small. The United States is to-day the world's greatest producer of manufactures, and as an exporter of domestic products it ranks side by side with Great Britain.

The Atlantic has given us near access to Europe, and the " American invasion " has followed. The Pacific has opened to us, though at longer range, the markets of the Orient, and the flag has been set up on an outlying fragment of the Asiatic continent. " Enthroned between her subject seas," the United States has by reason of her large area and her geographical location the most perfect conditions for attaining preëminence in the commerce of the world ocean.

If the tree planted by the fathers of the Republic has lifted its head high and spread its branches afar, due must be given to the great, generous land which has nourished and the seas which have watered its wide-running roots.

NOTES TO CHAPTER XIX

1. Mahan, The Problem of Asia, pp. 41, 65, 120. New York, 1900.
2. Çolquhoun, The Mastery of the Pacific, pp. 217–225. 1902.
3. H. Emerson, The Coal Resources of the Pacific Coasts, Engineering Magazine. May, 1902.
4. Colquhoun, The Mastery of the Pacific, p. 43. 1902.
5. Mahan, Interest of America in Sea Power, p. 48. 1897.
6. Annual Review of the Foreign Commerce of the United States for 1902, Table, p. 4342. Washington, 1902.
7. Ibid. Table, p. 4361.

INDEX

BIBLIOGRAPHY

Abert, Lieutenant, Examination of New Mexico, 1846–47. Executive Document no. 41. Washington, 1848.

American State Papers, Misc. vol. i. no. 250. Washington, 1834.

Bancroft, H. H., Arizona and New Mexico. 1889.

—— History of California. 1884.

—— History of the Northwest Coast.

Brigham, A. P., The Eastern Gateway of the United States. In the Geographical Journal, May, 1899, reprinted in The Journal of School Geography, April, 1900.

Brodhead, John Romeyn, History of the State of New York.

Brownell, History of Immigration.

Bruce, Philip Alexander, Economic History of Virginia in the Seventeenth Century.

Bulletin no. 12, Series 1901–1902, Export of Domestic Breadstuffs, etc. Bureau of Statistics, Washington, 1902.

Burke, Edmund, Speech on Conciliation.

Canadian Report of Immigration. Ottawa, 1901.

Census Reports. Eleventh Census of the United States taken in the year 1890.

Chittenden, H. M., The American Fur Trade of the Far West. New York, 1902.

Collins, R. H., History of Kentucky.

Colquhoun, Archibald Ross, The Mastery of the Pacific. 1902.

Cooke, Lieutenant-Colonel, Report of March from Santa Fé, New Mexico, to San Diego, California. Executive Document no. 41, Washington, 1848.

Coues, Elliott, History of the Lewis and Clark Expedition.

Davis, W. M., Physical Geography of Southern New England. Nat. Geog. Monographs, vol. i. no. 9.

Edward, David B., History of Texas ; or the Emigrant's, Farmer's, and Politician's Guide. Cincinnati, 1836.

Emerson, F. V., The Shenandoah Valley and the Civil War. Journal of School Geography, June, 1901.

Emerson, H., The Coal Resources of the Pacific Coasts. Engineering Magazine, May, 1902.

Emory, W. H., Notes of a Military Reconnoissance from Fort Leavenworth in Missouri to San Diego in California. Made in 1846–47. Executive Document no. 41, Washington, 1848.

—— Report of the United States and Mexico Boundary Commission. 1857.

Fairlie, J. A., Economic Effects of the Ship Canals. Annals of the American Academy, January, 1898.

Fiske, John, Discovery of America.

—— Mississippi Valley in the Civil War.

Flint, Timothy, History and Geography of the Mississippi Valley. 1832.

—— The Last Ten Years in the Valley of the Mississippi. Boston, 1826.

Ford, Paul Leicester, *Editor*. Writings of Thomas Jefferson.

Ford, Thomas, History of Illinois. Chicago, 1854.

Foster, John W., A Century of American Diplomacy. 1901.

Fremont, John Charles. Geographical Memoir of Upper California.

George, Hereford B., The Relations of Geography and History. Oxford, 1901.

Gregg, Josiah, The Commerce of the Prairies. 1845.

Hall, James, The West, its Commerce and Navigation. 1848.

Hart, Albert Bushnell, Foundation of American Foreign Policy. 1901.

Hartley, C. B., Life of Daniel Boone, Philadelphia, 1865.

Hayes, C. Willard, The Southern Appalachians. National Geographic Monographs, vol. i. no. 10.

Indians of Southwestern Alaska in Relation to their Environment. Journal of School Geography, June, 1898.

Ingersoll, Charles J., History of Second War of the United States with England. 1845.

Inman, Henry, the Old Santa Fé Trail. 1897.

Internal Commerce of the United States, Summary of Commerce and Finance. January, 1901.

Irving, Washington, Astoria. 1854.

Irving, Washington, Bonneville.

Key, C., Railway Development in Federated South Africa. Engineering Magazine, May, 1902.

Kirke, E., Chattanooga the Southern Gateway of the Alleghanies. Harper's Magazine, April, 1887.

Lewis, V. A., History of West Virginia. 1889.

Louisville : A Study in Economic Geography. Journal of School Geography, December, 1900.

Marshall, Humphrey, History of Kentucky. 1812.

McMaster, J. B., History of the People of the United States.

Mahan, A. T., Influence of Sea Power in History.

- —— The Interest of America in Sea Power.

—— The Problem of Asia. New York, 1900.

Monette, John W., History of the Valley of the Mississippi. 1846.

Morris, Henry C., History of Colonization from the earliest times to the present day. 1900.

Norway, Official Publication for the Paris Exposition. Christiania, 1900.

Parkman, Francis, Oregon Trail.

—— The Old Régime in Canada.

Peck, J. M., A Guide for Emigrants in Illinois and Missouri. Boston, 1831.

Pattie, James O. of Kentucky, Personal Narrative of. Cincinnati, 1831.

Poor's Manual of American Railroads, issues of 1880–1890.

Ratzel, Friedrich, Anthropogeographie.

—— Politische Geographie der Vereinigten Staaten. 1893.

—— Zur Kustenentwickelung. Jahresber. d. Geograph. Gesellschaft in Munich, 1894.

Report of the Isthmian Canal Commission, 1899 to 1901. Washington, 1901.

Report of Geological Survey, 1898, on " Waterway between Warrior River and Five-Mile Creek, Alabama."

Review of the World's Commerce for 1901. Washington, 1902.

Ripley, W. Z., The Races of Europe. 1899.

Roosevelt, Theodore, The Winning of the West.

Roscher, Wilhelm, Geschichte der National Oekonomik des Ackerbaues.

Russell, Israel, Rivers of North America.

Schaefer, Dietrich Die Hansestädte und König Waldemar von Dänemark. 1879.

Semple, Ellen Churchill, Development of the Hanse Towns in Relation to their Geographical Environment. Bulletin of the American Geographical Society, no. 3. 1899.

—— Mountain Passes : A Study in Anthropogeography. Bulletin of the American Geographical Society, nos. 2 and 3. 1901.

—— The Anglo-Saxons of the Kentucky Mountains. Geographical Journal, June, 1901.

Senate Report of Explorations and Surveys for a Railroad Route from the Mississippi River to the Pacific. Washington, 1861.

Shaler, Nathaniel Southgate, Kentucky. American Commonwealths Series.

—— Nature and Man in America.

Shaler, N. S., United States of America.

Smith, George Adam, Historical Geography of the Holy Land. 1897.

Smith, Richmond M., Emigration and Immigration. New York, 1890.

Sparks, Expansion of the American People.

Statistical Abstract of the United States for 1901, passim. Washington, 1902.

Summary of Commerce and Finance, August, 1901. Washington.

Treaties of the United States with the Several Indian Tribes, 1778–1837. Washington, 1837.

Turner, F. J., The Significance of the Frontier in American History. Annual Report of the American Historical Association for 1893.

—— Western State-Making in the Revolutionary Era. American Historical Review, vol. i. nos. 1 and 2.

Twelfth Census, Population.

United States Report of Commission of Navigation. 1901.

Volney, C. F., View of the Climate and Soil of the United States of America. London, 1804.

Walker, Dr. Thomas, Journal in Filson Club Publications, no. 13.

War of the Rebellion, Official Records.

Washington, George, Letter to Governor Harrison of Virginia.

—— Farewell Speech, September, 1796.

Weeden, W. B., Economic and Social History of New England, 1620–1789.

Whelpley, J. D., The Isolation of Canada, The Atlantic Monthly, August, 1901.

Willis, Bailey, The Northern Appalachians. National Geographic Monographs, vol. i. no. 6.

Winsor, Justin, The Westward Movement.

—— The Mississippi Basin.